Student Study Guide

for

The Mathematical Palette

Third Edition

Ronald Staszkow
Ohlone College

Robert Bradshaw
Ohlone College

THOMSON

BROOKS/COLE

Australia • Canada • Mexico • Singapore • Spain • United Kingdom • United States

For more information about our products, contact us at:
Thomson Learning Academic Resource Center
1-800-423-0563

For permission to use material from this text, contact us by:
Phone: 1-800-730-2214
Fax: 1-800-730-2215
Web: http://www.thomsonrights.com

Brooks/Cole—Thomson Learning, Inc.
10 Davis Drive
Belmont, CA 94002-3098
USA

Asia
Thomson Learning
5 Shenton Way #01-01
UIC Building
Singapore 068808

Australia/New Zealand
Thomson Learning
102 Dodds Street
Southbank, Victoria 3006
Australia

Canada
Nelson
1120 Birchmount Road
Toronto, Ontario M1K 5G4
Canada

Europe/Middle East/South Africa
Thomson Learning
High Holborn House
50/51 Bedford Row
London WC1R 4LR
United Kingdom

Latin America
Thomson Learning
Seneca, 53
Colonia Polanco
11560 Mexico D.F.
Mexico

Spain/Portugal
Paraninfo
Calle/Magallanes, 25
28015 Madrid, Spain

STUDENT STUDY GUIDE

PREFACE

The *Student Study Guide* for *The Mathematical Palette* 3/e, presents a brief summary of each section of the textbook. These summaries contain the important concepts, terms, and formulas used in each section. The *Student Study Guide* contains detailed solutions to the odd-numbered problems in each section and to all the problems in each chapter review and chapter test.

The *Student Study Guide* should be used along with *The Mathematical Palette 3/e*. It does not replace the textbook. It does not contain the depth of explanation, numerous examples, or summaries of terms and objectives found in the textbook.

It is our hope that the *Student Study Guide* will help you learn and understand the material contained in *The Mathematical Palette 3/e*.

CHAPTER 1 NUMBERS AND NUMERALS

The numerals 0 through 9 did not merely appear out of thin air. Even though they are generally accepted throughout the world, they were not the first numerals invented. In fact, many cultures have had entirely different systems of numeration, based on symbols that may seem entirely foreign to us. In this chapter, you will learn about different systems of numeration from the past and about different systems that are in use today. You will also learn about special types of numbers that have names of their own such as perfect numbers, friendly numbers, pentagonal numbers, binary, octal, and hexadecimal numbers. Finally, you will investigate number phenomena such as pi (π), Egyptian fractions, and real and imaginary numbers.

SECTION 1.1 ANCIENT SYSTEMS OF NUMERATION

Four types of ancient systems of numeration are discussed in this section. Each system contains symbols that represent tallies and methods for combining the symbols to represent numbers.

In **systems of numeration using addition**, the value of a numeral is found by simply adding the amounts represented by each symbol. The order in which each individual symbol is arranged does not change the value of the numeral. The Egyptian Hieroglyphic and Attic Greek systems are examples of additive systems of numeration.

In **systems of numeration using addition and subtraction**, numbers are represented by adding the value of each symbol. However, when writing numerals for certain numbers, subtraction is used. For example, the Roman numeral system the order in which symbols are placed indicates subtraction for numbers such as 4, 9, 40, 90, 400, 900, etc.

In **systems of numeration using addition and multiplication**, multiplication factors are placed near a symbol to indicate that the symbol is repeated. Symbols are not duplicated when many of the same symbols are needed. The traditional Chinese and Ionic Greek systems use this method in writing numerals.

Place value systems give a certain value to the position a symbol occupies in a numeral. The Babylonian system is based on 60, the Mayan system is based on 20 and 18, and the Hindu-Arabic system is based on 10.

A list of the symbols used in each ancient system is contained inside the back cover of the text book and in Section 1.1.

Explain

1. Systems that use addition require that the value of each symbol be added together to determine the value of the number. Examples include Egyptian Hieroglyphics and Attic Greek. A system that uses subtraction is one in which if a symbol representing a small value is written with symbol representing a larger value, the combined value is found by subtracting the smaller value from the larger value. For example, in Roman numerals XL represents the number 40 (50 – 10) while LX represents the number 60 (50 + 10).

3. To represent large numbers, systems that use addition and subtraction will generally need a larger number of symbols than a system using addition and multiplication.

5. A system that uses place value has the advantage over a system that uses addition and multiplication because only a small number of symbols must be memorized. However, a place value system requires a greater understanding of the mechanics of the number system.

7. Systems that do not use place value would require new symbols for larger and larger numbers. Thus, the Babylonian, Mayan, and Hindu-Arabic systems would not require new symbols while the other systems would require new symbols.

9. "For any number, there are many numerals." This means, for example, the number given by the tally "|||", can be represented in numerals in many ways, such as Hindu Arabic (3), Roman (III), and Ionic Greek (γ).

Apply

11. $100,000 + 20,000 + 1000 = 121,000$

13. $1,000,000 + 10,000 + 100 + 1 = 1,010,101$

15. $1000 + 100 + 5 + 1 = 1106$

17. $500 + 100 + 100 + 10 + 10 + 10 + 4 = 734$

19. $40,000 + 1000 + 1000 + 400 + 50 + 5 + 1 + 1 = 42,457$

21. $7 \times 1000 + 5 \times 100 + 2 = 7502$

23. $30 + 7 = 37$

25. $2 \times 10,000 + 5 \times 1000 + 60 + 2 = 20,000 + 5000 + 60 + 2 = 25,062$

27. $10 + 1 = 11$

29. $1 \times 360 + 0 + 3 = 363$

31. $19 \times 2,880,000 + 16 \times 144,000 + 6 \times 7200 + 4 \times 360 = 57,068,640$

33. (a) X   HHHH △△△△

    (b) —

        ╪

        ⬓

        百

        ⬓

        ╋

(c) ◄ ◄ ∀ ∀ ∀ ∀ ⌐

Explore

35. (a) X̄MMCCCXV

(b) —
     万

     =
     千

     ☰
     百

     —
     十

     五

(c) $\overset{\alpha}{M}$ '$\beta\tau\varepsilon$

(d) ∀ ∀ ∀   ◄ ◄ ∀ ∀ ∀ ∀ ∀   ◄ ∀ ∀ ∀ ∀ ∀

(e) ·
     ‗‗
     ···
     ≡

37. 1-800-YEA-MATH is the same as 1-800-932-6284. Answers will vary depending on the system of numeration chosen.

39. Additive systems are similar to the Attic Greek system. Answers will vary depending on the methods and symbols chosen. One possible solution is shown.
    (a) $75 = 8 \times 9 + 3 =$ ←←←←←←←← →→→
    (b) $366 = 4 \times 81 + 4 \times 9 + 6 =$ ↓↓↓↓ ←←←← →→→→→→
    (c) $5280 = 7 \times 729 + 2 \times 81 + 1 \times 9 + 6 =$ ↑↑↑↑↑↑↑ ↓↓ ← →→→→→→
    (d) $1000 = 1 \times 729 + 3 \times 81 + 3 \times 9 + 1 =$ ↑ ↓↓↓ ←←← →

41. Answers will vary depending on the methods and symbols chosen.

43. (a) In Egyptian, a carry takes place each time there are 10 of the same symbol. When this happens, an additional symbol of the next higher value will replace the group of ten symbols.

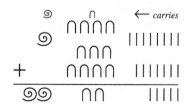

(b) In Roman, a carry takes place each time there are more than 4 of the symbols I, X, C, M or more than 1 of the symbols V, L, or D. When this happens, an additional symbol of the next higher value will replace the value that has been exceeded.

```
      C    L    X    V    ← carries
    C  L  XX   V   III
 +          XL   V   II
 ─────────────────────────
   CC       XX   V
```

(c) In Chinese, a carry takes place each time there are more than 9 in a particular place value. When this happens, 1 will be added to the next higher place value.

SECTION 1.2   HINDU-ARABIC SYSTEM AND FRACTIONS

The system of numeration used in the United States and in many other parts of the world is the Hindu-Arabic system. This decimal (base 10) system uses digits, 0, 1, 2, 3, 4, 5, 6, 7, 8, 9, and the powers of 10 to determine the place value of each digit used in a numeral.

To the left of the decimal point, the place values are determined by the non-negative integer powers of ten: 10^0 (ones place), 10^1 (tens place), 10^2 (hundreds place), 10^3 (thousands place), 10^4 (ten thousands place), 10^5 (hundred thousands place), 10^6 (millions place), etc.

To the right of the decimal point, the place values are determined by the negative integer powers of ten: 10^{-1} (tenths place), 10^{-2} (hundredths place), 10^{-3} (thousandths place), 10^{-4} (ten-thousandths place), 10^{-5} (hundred-thousandths place), 10^{-6} (millionths place), etc. The Hindu-Arabic numeral 5763.24 can be represented by powers of ten as shown below:

$$5763.24 = 5 \times 10^3 + 7 \times 10^2 + 6 \times 10^1 + 3 \times 10^0 + 2 \times 10^{-1} + 4 \times 10^{-2}$$

Fractional numbers were also represented in ancient systems of numeration. The Egyptians, Babylonians, Romans, and Greeks had interesting schemes for representing fractions. Section 1.2 contains the details on how these systems represented fractions.

Explain

1. A decimal fraction is a number such as 0.3 while a fraction is a number such as 2/5.

3. In the expanded form of a Hindu Arabic numeral, the number is written in terms of the powers of ten. For example, $345 = 3 \times 10^2 + 4 \times 10^1 + 5 \times 10^0$.

5. The Babylonian system of fractions is similar to the decimal system of fractions in that both have a symbol to indicate fractions (𝄈 and decimal point), both use place value, and both allow numerators greater than one.

Apply

7. $1 \times 10^2 + 3 \times 10^1 + 9 \times 10^0$

9. $4 \times 10^2 + 3 \times 10^1 + 7 \times 10^0 + 1 \times 10^{-1} + 5 \times 10^{-2}$

11. $3 \times 10^{-1} + 1 \times 10^{-2} + 4 \times 10^{-3}$

13. $5 \times 10^5 + 4 \times 10^4 + 3 \times 10^3 + 8 \times 10^2 + 6 \times 10^1 + 7 \times 10^0$

15. $5 \times 10^1 + 3 \times 10^0 + 1 \times 10^{-1} + 7 \times 10^{-2} + 1 \times 10^{-3}$

17. $6 \times 10^{-1} + 2 \times 10^{-2} + 1 \times 10^{-3} + 9 \times 10^{-4} + 3 \times 10^{-5}$

19. tens, hundredths, tens, ten thousands, tenths, hundredths, thousands, hundredths, ones, ones, hundred thousandths, thousandths

21. 1/10 + 1/4 = 7/20

23. 1/1000 + 1/300 + 1/20 = 163/3000

25. 12/60 = 1/5

27. 10/60 + 10/3600 + 10/216,000 = 3661/21,600

29. 4/12 = 1/3

31. 9/12 = 3/4

33. 1/63

35. The numerator $\mu\theta = 49$. The denominator $\phi\alpha = 501$. Thus, $\mu\theta$ $\phi\alpha'$ $\phi\alpha' = 49/501$.

Explore

37. (12)(60) + 23 + 12/60 = 743 1/5

39. 101 1/34

41. $5/12 = 4/12 + 1/12 = 1/3 + 1/12 =$ ☺   ☺
                                             ||| ∩||

   or $5/12 = 2/12 + 3/12 = 1/6 + 1/4 =$ ☺   ☺
                                            ||||| ||||

43. $8/15 = 5/15 + 3/15 = 1/3 + 1/5 =$ ☺   ☺
                                          ||| |||||

45.   ☺ ☺
     ||| ||

47. . . . . + S = S . . . .

49. XII × III S... = XLV

51. $\rho\kappa \times \gamma \; \gamma\delta'\delta' = \upsilon\nu$

53. Answers will vary.

55. Answers will vary.

SECTION 1.3   NUMERATION SYSTEMS WITH OTHER BASES

While the Hindu-Arabic system uses the powers of 10 to determine the place-value of each digit contained in a numeral, other place-value systems can be created using any whole number greater than one as its base. The digits of such a place-value system will consist of all the whole numbers less than the base. If b is the base of a place-value system, the place value of each digit to the left of the basimal point (dot separating whole and fractional parts) is determined by non-negative integer powers of the base: $b^0, b^1, b^2, b^3, b^4, b^5$, etc. The place value of each digit to the right of the basimal point is determined by negative integer powers of the base: $b^{-1}, b^{-2}, b^{-3}, b^{-4}, b^{-5}$, etc. For example, by examining the place-value structure of the numeral 573.4 in base eight, we can see that it is equivalent to the Hindu-Arabic numeral 379.5.

$$573.4_8$$

$$4 \times 8^{-1} = 4 \times 0.125 = \quad 0.5$$
$$3 \times 8^0 = 3 \times 1 \quad = \quad 3$$
$$7 \times 8^1 = 7 \times 8 \quad = \quad 56$$
$$5 \times 8^2 = 5 \times 64 \quad = \underline{320}$$
$$379.5$$

To convert a base ten numeral to another base you must find how many groups of each appropriate place-value is contained in the base ten numeral. The following example shows how this is done.

Convert 210_{10} to base 4.

The place values of a base 4 system that are less than 210 are $4^3 = 64$, $4^2 = 16$, $4^1 = 4$, and $4^0 = 1$. Using a systematic process of division, the number of times each of these place-values is contained in 210 can be determined.

$$
\begin{array}{cccc}
3 & 1 & 0 & 2 \\
64\overline{)210} & 16\overline{)18} & 4\overline{)2} & 1\overline{)2} \\
\underline{192} & \underline{16} & \underline{0} & \underline{2} \\
18 & 2 & 2 & 0
\end{array}
$$

Thus, 210 contains 3 groups of 64, 1 group of 16, 0 groups of 4, and 2 groups of 1.

$210_{10} = 3 \times 64 + 1 \times 16 + 0 \times 4 + 2 \times 1 = 3 \times 4^3 + 1 \times 4^2 + 0 \times 4^1 + 2 \times 4^0 = 3102_4$

Explain

1. Ten symbols and a decimal point are used to create a number in base ten.

3. It is thought that modern cultures use base 10 because humans have 10 fingers.

5. To convert 12_8 into base 10, recognize that the 1 represents one group of 8^1 while the 2 represents two groups of 8^0 or (1). Therefore $12_8 = 1 \times 8^1 + 2 \times 8^0 = 10$.

Apply

7. 77
$$302_5 = 3 \times 5^2 + 0 \times 5^1 + 2 \times 5^0 = 3 \times 25 + 0 + 2 \times 1 = 77$$

9. 5.625
$$
\begin{aligned}
101.101_2 &= 1 \times 2^2 + 0 \times 2^1 + 1 \times 2^0 + 1 \times 2^{-1} + 0 \times 2^{-2} + 1 \times 2^{-3} \\
&= 1 \times 4 + 0 \times 2 + 1 \times 2^0 + 1 \div 2 + 0 \div 4 + 1 \div 8 \\
&= 5.625
\end{aligned}
$$

11. 2952
$$5610_8 = 5 \times 8^3 + 6 \times 8^2 + 1 \times 8^1 + 0 \times 8^0 = 5 \times 512 + 6 \times 64 + 1 \times 8 + 0 = 2952$$

13. 1132.25
$$
\begin{aligned}
7A4.3_{12} &= 7 \times 12^2 + 10 \times 12^1 + 4 \times 1 + 3 \times 12^{-1} \\
&= 7 \times 144 + 10 \times 12 + 4 + 3 \div 12 \\
&= 1132.25
\end{aligned}
$$

15. 136
$$253_7 = 2 \times 7^2 + 5 \times 7^1 + 3 \times 7^0 = 2 \times 49 + 5 \times 7 + 3 = 136$$

17. 15

$$1111_2 = 1 \times 2^3 + 1 \times 2^2 + 1 \times 2^1 + 1 \times 2^0 = 8 + 4 + 2 + 1 = 15$$

19. 3101_5

```
        3          1          0          1
125)401     25)26       5)1        1)1
    375         25         0          1
    ----        --         -          -
     26          1         1          0
```

21. 1736_8

```
        1          7          3          6
512)990     64)478      8)30       1)6
    512         448         24         6
    ----        ----        --         -
    478          30          6         0
```

23. 2160_7

```
        2          1          6          0
343)777     49)91       7)42       1)0
    686         49          42         0
    ----        --          --         -
     91         42           0         0
```

25. $10B0_{12}$

```
         1          0           11          0
1728)1860   144)132      12)132       1)0
     1728        0            132          0
     ----        ---          ---          -
      132        132            0          0
```

27. 81_{16}

```
        8          1
16)129      1)1
   128         1
   ---         -
     1         0
```

29. 313 126₉

```
            3          1          3          1          2          6
59,049)186,000  6561)8853  729)2292   81)105      9)24       1)6
       177,147       6561      2187       81         18         6
       -------       ----      ----       --         --         -
         8,853       2292       105       24          6         0
```

31. 11 011 110₂

```
        1          1          0          1          1          1          1          0
128)222     64)94      32)30      16)30       8)14       4)6        2)2        1)0
    128         64          0         16          8          4          2          0
    ---         --          -         --          -          -          -          -
     94         30         30         14          6          2          0          0
```

33. 2431_5

$$\begin{array}{r}2\\125\overline{)366}\\\underline{250}\\116\end{array}\qquad\begin{array}{r}4\\25\overline{)116}\\\underline{100}\\16\end{array}\qquad\begin{array}{r}3\\5\overline{)16}\\\underline{15}\\1\end{array}\qquad\begin{array}{r}1\\1\overline{)1}\\\underline{1}\\0\end{array}$$

35. 556_8

$$\begin{array}{r}5\\64\overline{)366}\\\underline{320}\\46\end{array}\qquad\begin{array}{r}5\\8\overline{)46}\\\underline{40}\\6\end{array}\qquad\begin{array}{r}6\\1\overline{)6}\\\underline{6}\\0\end{array}$$

37. $7D0_{16}$

$$\begin{array}{r}7\\256\overline{)2000}\\\underline{1792}\\208\end{array}\qquad\begin{array}{r}13\\16\overline{)208}\\\underline{208}\\0\end{array}\qquad\begin{array}{r}0\\1\overline{)0}\\\underline{0}\\0\end{array}$$

39. $224\,000\,000_5$

$$\begin{array}{r}2\\390{,}625\overline{)1{,}000{,}000}\\\underline{781{,}250}\\218{,}750\end{array}\qquad\begin{array}{r}2\\78{,}125\overline{)218{,}750}\\\underline{156{,}250}\\62{,}500\end{array}\qquad\begin{array}{r}4\\15{,}625\overline{)62{,}500}\\\underline{62{,}500}\\0\end{array}$$

41. $3\,641\,100_8$

$$\begin{array}{r}3\\262{,}144\overline{)1{,}000{,}000}\\\underline{786{,}432}\\213{,}568\end{array}\quad\begin{array}{r}6\\32{,}768\overline{)213{,}568}\\\underline{196{,}608}\\16{,}960\end{array}\quad\begin{array}{r}4\\4096\overline{)16{,}960}\\\underline{16{,}384}\\576\end{array}\quad\begin{array}{r}1\\512\overline{)576}\\\underline{512}\\64\end{array}\quad\begin{array}{r}1\\64\overline{)64}\\\underline{64}\\0\end{array}\quad\begin{array}{r}0\\8\overline{)0}\\\underline{0}\\0\end{array}\quad\begin{array}{r}0\\1\overline{)0}\\\underline{0}\\0\end{array}$$

43. $F4240_{16}$

$$\begin{array}{r}15\\65{,}536\overline{)1{,}000{,}000}\\\underline{983{,}040}\\16{,}960\end{array}\quad\begin{array}{r}4\\4096\overline{)16{,}960}\\\underline{16{,}384}\\576\end{array}\quad\begin{array}{r}2\\256\overline{)576}\\\underline{512}\\64\end{array}\quad\begin{array}{r}4\\16\overline{)64}\\\underline{64}\\0\end{array}\quad\begin{array}{r}0\\1\overline{)0}\\\underline{0}\\0\end{array}$$

Explore

45. 254_9

254_9 is larger since $254_9 = 2 \times 9^2 + 5 \times 9 + 4 = 211$ while
$12202_3 = 1 \times 3^4 + 2 \times 3^3 + 2 \times 3^2 + 2 = 155$.

47. 3033_4

3033_4 is larger since $101\,101_2 = 1 \times 2^5 + 1 \times 2^3 + 1 \times 2^2 + 1 = 45$ while
$3033_4 = 3 \times 4^3 + 3 \times 4 + 3 = 207$.

49. 150_7

Since $45_6 = 4 \times 6 + 5 = 29$ and $67_8 = 6 \times 8 + 7 = 55$, $45_6 + 67_8 = 29 + 55 = 84$
84 in base 7 is 150_7.

$$
\begin{array}{r} 1 \\ 49{\overline{)84}} \\ \underline{49} \\ 35 \end{array}
\quad
\begin{array}{r} 5 \\ 7{\overline{)35}} \\ \underline{35} \\ 0 \end{array}
\quad
\begin{array}{r} 0 \\ 1{\overline{)0}} \\ \underline{0} \\ 0 \end{array}
$$

Thus, $45_6 + 67_8 = 150_7$

51. $B = 8$

$$200_B = 2 \times B^2 \text{ so } 2B^2 = 128$$
$$B^2 = 64$$
$$B = 8$$

53. (a) $45 = 1200_3$

$$
\begin{array}{r} 1 \\ 27{\overline{)45}} \\ \underline{27} \\ 18 \end{array}
\quad
\begin{array}{r} 2 \\ 9{\overline{)18}} \\ \underline{18} \\ 0 \end{array}
\quad
\begin{array}{r} 0 \\ 3{\overline{)0}} \\ \underline{0} \\ 0 \end{array}
\quad
\begin{array}{r} 0 \\ 1{\overline{)0}} \\ \underline{0} \\ 0 \end{array}
$$

$100 = 10201_3$

$$
\begin{array}{r} 1 \\ 81{\overline{)100}} \\ \underline{81} \\ 19 \end{array}
\quad
\begin{array}{r} 0 \\ 27{\overline{)19}} \\ \underline{0} \\ 19 \end{array}
\quad
\begin{array}{r} 2 \\ 9{\overline{)19}} \\ \underline{18} \\ 1 \end{array}
\quad
\begin{array}{r} 0 \\ 3{\overline{)1}} \\ \underline{0} \\ 1 \end{array}
\quad
\begin{array}{r} 1 \\ 1{\overline{)1}} \\ \underline{1} \\ 0 \end{array}
$$

$200 = 21102_3$

$$
\begin{array}{r} 2 \\ 81{\overline{)200}} \\ \underline{162} \\ 38 \end{array}
\quad
\begin{array}{r} 1 \\ 27{\overline{)38}} \\ \underline{27} \\ 11 \end{array}
\quad
\begin{array}{r} 1 \\ 9{\overline{)11}} \\ \underline{9} \\ 2 \end{array}
\quad
\begin{array}{r} 0 \\ 3{\overline{)2}} \\ \underline{0} \\ 2 \end{array}
\quad
\begin{array}{r} 2 \\ 1{\overline{)2}} \\ \underline{2} \\ 0 \end{array}
$$

$45 = 50_9$

$$
\begin{array}{r} 5 \\ 9{\overline{)45}} \\ \underline{45} \\ 0 \end{array}
\quad
\begin{array}{r} 0 \\ 1{\overline{)0}} \\ \underline{0} \\ 0 \end{array}
$$

$100 = 121_9$

$$
\begin{array}{r} 1 \\ 81{\overline{)100}} \\ \underline{81} \\ 19 \end{array}
\quad
\begin{array}{r} 2 \\ 9{\overline{)19}} \\ \underline{18} \\ 1 \end{array}
\quad
\begin{array}{r} 1 \\ 1{\overline{)1}} \\ \underline{1} \\ 0 \end{array}
$$

$200 = 242_9$

$$
\begin{array}{r} 2 \\ 81{\overline{)200}} \\ \underline{162} \\ 38 \end{array}
\quad
\begin{array}{r} 4 \\ 9{\overline{)38}} \\ \underline{36} \\ 2 \end{array}
\quad
\begin{array}{r} 2 \\ 1{\overline{)2}} \\ \underline{2} \\ 0 \end{array}
$$

(b) As the base increases, the number of digits decreases.

(c) Taken in pairs, the digits in the base 3 representation give the base 9 representation. For example, writing 21102_3 as 2_3 11_3 02_3, we have 2, 4, and 2. These are the digits in the base 9 representation of 200.

55. Answers will vary.

57. When using a base greater than 10, the telephone number will contain the same number of digits if you consider each digit separately. If you consider the digits as three and four digit numbers rather than separate digits, the new number will contain the same or fewer digits.

SECTION 1.4   THE NUMBERS OF TECHNOLOGY

Section 1.4 introduces the numbers of the computer age. It begins with an explanation of how data is read and stored using the binary digits – 1 and 0. It discusses how information is processed using ASCII (American Standard Code for Information Exchange) and how these codes can be converted to octal and hexadecimal numerals. The section ends by giving some insight into how computation is accomplished using binary numbers.

Explain

1. The digits 0 and 1 represent electric pulses having low and high voltage levels.

3. ASCII stands for American Standard Code for Information Exchange. Computers use ASCII to send and receive information.

5. $1 + 1 = 2$ and $10_2 = 1 \times 2^1 + 0 \times 2^0 = 1 \times 2 + 0 \times 1 = 2 + 0 = 2$. Thus, $1 + 1 = 10_2$.

7. In order to multiply numbers, the computer converts decimal numerals into binary and multiplication is done electronically using the binary multiplication facts, $0 \times 0 = 0$, $0 \times 1 = 0$, $1 \times 0 = 0$, and $1 \times 1 = 1$.

Apply

9. (a) 25_8

Separating the binary numeral into groups of three, and finding the value of each group using the place values of 2^2, 2^1, and 2^0 (4, 2, 1), we get the following.

$$\underbrace{\overset{2\ 1}{1\,0}}_{2}\ \underbrace{\overset{4\ 2\ 1}{1\,0\,1}}_{5} = 25_8$$

(b) 15_{16}

Separating the binary numeral into groups of four, and finding the value of each group using the place values of 2^3, 2^2, 2^1, and 2^0 (8, 4, 2, 1), we get the following.

$$\underbrace{\overset{1}{1}}_{1}\ \underbrace{\overset{8\ 4\ 2\ 1}{0\,1\,0\,1}}_{5} = 15_{16}$$

11. (a) 144_8

Separating the binary numeral into groups of three, and finding the value of each group using the place values of 2^2, 2^1, and 2^0 (4, 2, 1), we get the following.

$$\underset{1}{\underbrace{\overset{1}{1}}}\ \underset{4}{\underbrace{\overset{4\ 2\ 1}{1\ 0\ 0}}}\ \underset{4}{\underbrace{\overset{4\ 2\ 1}{1\ 0\ 0}}} = 144_8$$

(b) 64_{16}

Separating the binary numeral into groups of four, and finding the value of each group using the place values of 2^3, 2^2, 2^1, and 2^0 (8, 4, 2, 1), we get the following.

$$\underset{6}{\underbrace{\overset{4\ 2\ 1}{1\ 1\ 0}}}\ \underset{4}{\underbrace{\overset{8\ 4\ 2\ 1}{0\ 1\ 0\ 0}}} = 64_{16}$$

13. (a) 5656_8

Separating the binary numeral into groups of three, and finding the value of each group using the place values of 2^2, 2^1, and 2^0 (4, 2, 1), we get the following.

$$\underset{5}{\underbrace{\overset{4\ 2\ 1}{1\ 0\ 1}}}\ \underset{6}{\underbrace{\overset{4\ 2\ 1}{1\ 1\ 0}}}\ \underset{5}{\underbrace{\overset{4\ 2\ 1}{1\ 0\ 1}}}\ \underset{6}{\underbrace{\overset{4\ 2\ 1}{1\ 1\ 0}}} = 5656_8$$

(b) BAE_{16}

Separating the binary numeral into groups of four, and finding the value of each group using the place values of 2^3, 2^2, 2^1, and 2^0 (8, 4, 2, 1), we get the following.

$$\underset{\underset{B}{11}}{\underbrace{\overset{8\ 4\ 2\ 1}{1\ 0\ 1\ 1}}}\ \underset{\underset{A}{10}}{\underbrace{\overset{8\ 4\ 2\ 1}{1\ 0\ 1\ 0}}}\ \underset{\underset{E}{14}}{\underbrace{\overset{8\ 4\ 2\ 1}{1\ 1\ 1\ 0}}} = BAE_{16}$$

15. $0110\ 0101_2$

To convert the hexadecimal ASCII code into binary, find the four digit binary code for each of the digits of the hexadecimal numeral using the place values of 2^3, 2^2, 2^1, and 2^0 (8, 4, 2, 1).

$$\overset{6}{\overbrace{110}}\ \overset{5}{\overbrace{0101}} = 0110\ 0101_2$$

17. $0010\ 0101_2$

To convert the hexadecimal ASCII code into binary, find the four digit binary code for each of the digits of the hexadecimal numeral using the place values of 2^3, 2^2, 2^1, and 2^0 (8, 4, 2, 1).

$$\overset{2}{\overbrace{10}}\ \overset{5}{\overbrace{0101}} = 0010\ 0101_2$$

19. $0010\ 1011_2$

   To convert the hexadecimal ASCII code into binary, find the four digit binary code for each of the digits of the hexadecimal numeral using the place values of 2^3, 2^2, 2^1, and 2^0 (8, 4, 2, 1).

$$\underbrace{2}_{10}\ \underbrace{B}_{1011} = 0010\ 1011_2$$

21. $1\ 001\ 001_2$

$$
\begin{array}{r}
\overset{1\ \ 1\ \ 1}{110\ \ 011} \\
+\ \ \ 10\ \ 110 \\
\hline
1\ 001\ \ 001
\end{array}
$$

23. $1\ 100\ 011\ 110_2$

$$
\begin{array}{r}
\overset{11\qquad\ \ 1}{111\ 000\ 101} \\
+\ \ 101\ 011\ 001 \\
\hline
1\ 100\ 011\ \ 110
\end{array}
$$

25. $100\ 000\ 100_2$

$$
\begin{array}{r}
10\ 100 \\
\times\ \ 1\ 101 \\
\hline
10\ 100 \\
000\ 00 \\
1\ 010\ 0 \\
10\ 100 \\
\hline
100\ 000\ 100
\end{array}
$$

27. $110\ 001\ 110\ 001_2$

$$
\begin{array}{r}
111\ 000\ 111 \\
\times\qquad\quad 111 \\
\hline
111\ 000\ 111 \\
1\ 110\ 001\ 11 \\
11\ 100\ 011\ 1 \\
\hline
110\ 001\ 110\ 001
\end{array}
$$

Explore

29. To be or not?
    The binary codes converted to hexadecimal are: 54, 6F, 20, 62, 65, 20, 6F, 72, 20, 6E, 6F, 74, 3F. From the ASCII chart we get: To be or not?

31. 0100 1000, 0100 1111, 0101 0000, 0100 0101

    $H = 48_{16} = 0100\ 1000_2$
    $O = 4F_{16} = 0100\ 1111_2$
    $P = 50_{16} = 0101\ 0000_2$
    $E = 45_{16} = 0100\ 0101_2$

33. 0100 1101, 0110 1111, 0110 1101

    $M = 4D_{16} = 0100\ 1101_2$
    $o = 6F_{16} = 0110\ 1111_2$
    $m = 6D_{16} = 0110\ 1101_2$

35. Divide the binary number in groups of five digits. Using the place value for five binary digits, 16, 8, 4, 2, 1, find the value of each group of 5 digits. Remember to use A for 10, B for 11, C for 12, ..., W for 32.

SECTION 1.5 TYPES OF NUMBERS

This section presents many different types of numbers. It contains a review of real and complex numbers, an introduction to numbers based on factors and geometric shapes, and historical notes on zero, pi, negative numbers, and prime numbers. The section also demonstrates some of the historical interest in numerical patterns by introducing perfect numbers and amicable numbers.

Explain

1. Prime numbers are natural numbers whose only proper factor is 1.

3. An abundant number is a natural number with the property that the sum of its proper factors is more than the number.

5. Since real numbers can be either rational or irrational, it is possible to have a real number that is not rational. An example is π.

7. Zero is used in numeration systems that use place values. As a result, it was used by the Babylonians and the Mayans. The symbol for zero is believed to have developed in the 9th century in the Gwalior region of India or in the 7th century in Indochina. It became well established in Europe in the late 1400's. In western cultures, negative numbers were first used as solutions to equations by the Greeks in 270 B.C. but were dismissed as being absurd. It was not until the 16th century that negative numbers gained general acceptance. In non-western cultures such as China and India, negative numbers were well accepted as early as the 2nd century B.C.

9. Even numbers are whole numbers that have 2 as a proper factor.

11. Every prime number has 1 as its only proper factor, thus the sum of the proper factors of every prime number is 1. Since the first prime number is 2, every prime is greater than the sum of its proper factors. Since this is the definition of a deficient number, every prime must be deficient.

Apply

13. (a) $\sqrt{100}$, 19
    (b) 0, $\sqrt{100}$, 19
    (c) -7, $-\sqrt{64}$, 0, $\sqrt{100}$, 19
    (d) -14.785, -7, $-\sqrt{64}$, $\dfrac{-5}{16}$, 0, $5\dfrac{7}{8}$, 9.76555..., $\sqrt{100}$, 19
    (e) $\sqrt[3]{19}$, $\sqrt{8}$, π
    (f) -14.785, -7, $-\sqrt{64}$, $\dfrac{-5}{16}$, 0, $\sqrt[3]{19}$, $\sqrt{8}$, π, $5\dfrac{7}{8}$, 9.76555..., $\sqrt{100}$, 19
    (g) $-\sqrt{-25}$, $\sqrt{-8}$
    (h) π

15.

 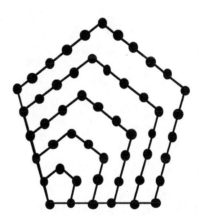

The fifth pentagonal number is 35.   The sixth pentagonal number is 51.

17. (a) prime, deficient
    (b) composite, deficient - proper factors are 1, 7, 11
    (c) composite, deficient - proper factors are 1, 5, 29
    (d) composite, abundant - proper factors are 1, 2, 4, 7, 14, 28, 71, 142, 284, 497, 994
    (e) composite, perfect - proper factors are 1, 2, 4, 8, 16, 32, 64, 127, 254, 508, 1016, 2032, 4064
    (f) composite, abundant - proper factors are 1, 2, 3, 4, 5, 6, 8, 9, 10, 12, 15, 18, 20, 24, 25, 30, 36, 40, 45, 50, 60, 72, 75, 90, 100, 120, 125, 150, 180, 200, 225, 250, 300, 360, 375, 450, 500, 600, 750, 900, 1000, 1125, 1500, 1800, 2250, 3000, 4500

Explore

19. Pattern: add the odd numbers (3, 5, 7, 9, 11, ...) to proceed from term to term.
    1, 4, 9, 16, 25, 36, 49, 64, 81, 100, 121, 144

21. The number 5 is a prime number. However, all others numbers ending in 5 are divisible by 5, so they are not prime.

23. 1.414 is a rational number and is approximately equal to $\sqrt{2}$ but $\sqrt{2}$ is an irrational number. The correct relationship is $\sqrt{2} > 1.414$.

CHAPTER 1 REVIEW

Review Section 1.1

1. (a) 123

    (b) ⊙∩∩|||

    (c) CXXIII

    (d) —
         ⌂

         =

         +

         ☰

    (e) ρκγ

    (f) HΔΔ|||

    (g) 123 = 2(60) + 3 = ❤❤   ❤❤❤

    (h) ∴
         •••

2. (a) 14   (b) 5614   (c) 8945   (e) 25,560   (f) 10,403

Review Section 1.2

3. (a) $2 \times 10^1 + 5 \times 10^0 + 4 \times 10^{-1} + 7 \times 10^{-2}$
    (b) $1 \times 10^2 + 2 \times 10^1 + 0 \times 10^0 + 0 \times 10^{-1} + 4 \times 10^{-2} + 5 \times 10^{-3}$

4. (a) 0.75

    (b) S . . .

(c)   ◉ ◉
       | | | | | |

(d)   ⟨⟨ ◄ ◄ ◄ ◄ ∨ ∨ ∨ ∨ ∨

(e) $\gamma\delta'\delta'$

5.   Answers will vary depending on what is selected.

Review Section 1.3

6.   (a) 194

$$1234_5 = 1 \times 5^3 + 2 \times 5^2 + 3 \times 5^1 + 4 \times 5^0$$
$$= 1 \times 125 + 2 \times 25 + 3 \times 5 + 4 \times 1$$
$$= 125 + 50 + 15 + 4$$
$$= 194$$

(b) 10 7/16 = 10.4375

$$12.34_8 = 1 \times 8^1 + 2 \times 8^0 + 3 \times 8^{-1} + 4 \times 8^{-2}$$
$$= 8 + 2 + \frac{3}{8} + \frac{4}{64}$$
$$= 10 + \frac{6}{16} + \frac{1}{16}$$
$$= 10\frac{7}{16} = 10.4375$$

(c) 2140

$$12A4_{12} = 1 \times 12^3 + 2 \times 12^2 + 10 \times 12^1 + 4 \times 12^0$$
$$= 1 \times 1728 + 2 \times 144 + 10 \times 12 + 4 \times 1$$
$$= 1728 + 288 + 120 + 4$$
$$= 2140$$

(d) 194 1/64 = 194.015625

$$C2.04_{16} = 12 \times 16^1 + 2 \times 16^0 + 0 \times 16^{-1} + 4 \times 16^{-2}$$
$$= 192 + 2 + \frac{4}{256}$$
$$= 194\frac{1}{64} = 194.015625$$

7. (a) $1\ 111\ 011_2$

$$
\begin{array}{c}
1 \\
64)\overline{123} \\
\underline{64} \\
59
\end{array}
\quad
\begin{array}{c}
1 \\
32)\overline{59} \\
\underline{32} \\
27
\end{array}
\quad
\begin{array}{c}
1 \\
16)\overline{27} \\
\underline{16} \\
11
\end{array}
\quad
\begin{array}{c}
1 \\
8)\overline{11} \\
\underline{8} \\
3
\end{array}
\quad
\begin{array}{c}
0 \\
4)\overline{3} \\
\underline{0} \\
3
\end{array}
\quad
\begin{array}{c}
1 \\
2)\overline{3} \\
\underline{2} \\
1
\end{array}
\quad
\begin{array}{c}
1 \\
1)\overline{1} \\
\underline{1} \\
0
\end{array}
$$

(b) 443_5

$$
\begin{array}{c}
4 \\
25)\overline{123} \\
\underline{100} \\
23
\end{array}
\quad
\begin{array}{c}
4 \\
5)\overline{23} \\
\underline{20} \\
3
\end{array}
\quad
\begin{array}{c}
3 \\
1)\overline{3} \\
\underline{3} \\
0
\end{array}
$$

(c) 173_8

$$
\begin{array}{c}
1 \\
64)\overline{123} \\
\underline{64} \\
59
\end{array}
\quad
\begin{array}{c}
7 \\
8)\overline{59} \\
\underline{56} \\
3
\end{array}
\quad
\begin{array}{c}
3 \\
1)\overline{3} \\
\underline{3} \\
0
\end{array}
$$

(d) $A3_{12}$

$$
\begin{array}{c}
10 \\
12)\overline{123} \\
\underline{120} \\
3
\end{array}
\quad
\begin{array}{c}
3 \\
1)\overline{3} \\
\underline{3} \\
0
\end{array}
$$

(e) $7B_{16}$

$$
\begin{array}{c}
7 \\
16)\overline{123} \\
\underline{112} \\
11
\end{array}
\quad
\begin{array}{c}
11 \\
1)\overline{11} \\
\underline{11} \\
0
\end{array}
$$

Review Section 1.4

8. Disks, tapes, and CDs store data using two different states – low/high voltage pulses, forward/backward magnetic fields, and dark/light marks. These two states are associated with the 0's and 1's of the binary numbers system.

9. 0100 0010, 0100 0100, 0110 1111, 0010 0000, 0110 1001, 0111 0100, 0010 0001, 0100 0010

10. 563_8, 173_{16}

$$
\underbrace{1\ 0\ 1}_{\substack{4+0+1 \\ 5}}\ \underbrace{1\ 1\ 0}_{\substack{4+2+0 \\ 6}}\ \underbrace{0\ 1\ 1}_{\substack{0+2+1 \\ 3}} = 563_8
\qquad
\underbrace{1}_{\substack{1 \\ 1}}\ \underbrace{0\ 1\ 1\ 1}_{\substack{0+4+2+1 \\ 7}}\ \underbrace{0\ 0\ 1\ 1}_{\substack{0+0+2+1 \\ 3}} = 173_{16}
$$

11. $1\ 000\ 011_2,\ 103_8,\ 43_{16}$

$$64\overline{)67} \quad 32\overline{)3} \quad 16\overline{)3} \quad 8\overline{)3} \quad 4\overline{)3} \quad 2\overline{)3} \quad 1\overline{)1} = 1\ 000\ 011_2$$

$$
\begin{array}{llllllll}
 & 1 & 0 & 0 & 0 & 0 & 1 & 1 \\
64\overline{)67} & 32\overline{)3} & 16\overline{)3} & 8\overline{)3} & 4\overline{)3} & 2\overline{)3} & 1\overline{)1} \\
\underline{64} & \underline{0} & \underline{0} & \underline{0} & \underline{0} & \underline{2} & \underline{1} \\
3 & 3 & 3 & 3 & 3 & 1 & 0
\end{array} = 1\ 000\ 011_2
$$

$$
\begin{array}{lll}
 & 1 & 0 & 3 \\
64\overline{)67} & 8\overline{)3} & 1\overline{)3} \\
\underline{64} & \underline{0} & \underline{3} \\
3 & 3 & 0
\end{array} = 103_8
$$

$$
\begin{array}{ll}
 & 4 & 3 \\
16\overline{)67} & 1\overline{)3} \\
\underline{64} & \underline{3} \\
3 & 0
\end{array} = 43_{16}
$$

12. (a) $111\ 011\ 010_2$

$$
\begin{array}{r}
\overset{1\,1\quad\ 1\ 1\,1}{101\ 110\ 011} \\
+\quad 1\ 100\ 111 \\
\hline
111\ 011\ 010
\end{array}
$$

$101110011_2 = 1\times2^8 + 0\times2^7 + 1\times2^6 + 1\times2^5 + 1\times2^4 + 0\times2^3 + 0\times2^2 + 1\times2^1 + 1\times2^0$ å

$\qquad = 256 + 64 + 32 + 16 + 2 + 1$

$\qquad = 371$

$1100111_2 = 1\times2^6 + 1\times2^5 + 0\times2^4 + 0\times2^3 + 1\times2^2 + 1\times2^1 + 1\times2^0$

$\qquad = 64 + 32 + 4 + 2 + 1$

$\qquad = 103$

$111011010_2 = 1\times2^8 + 1\times2^7 + 1\times2^6 + 0\times2^5 + 1\times2^4 + 1\times2^3 + 0\times2^2 + 1\times2^1 + 0\times2^0$

$\qquad = 256 + 128 + 64 + 16 + 8 + 2$

$\qquad = 474$

Check: $371 + 103 = 474$

(b) $11\,100\,111_2$

```
            1 0 1 0 1
     ×        1 0 1 1
   ─────────────────────
           ¹1 0 1 0 1
         ¹1  0 1 0 1
       ¹0  0  0 0 0
     1  0  1  0 1
   ─────────────────────
     1  1  1  0 0 1 1 1
```

$$10101_2 = 1\times2^4 + 0\times2^3 + 1\times2^2 + 0\times2^1 + 1\times2^0$$
$$= 16+4+1$$
$$= 21$$

$$1011_2 = 1\times2^3 + 0\times2^2 + 1\times2^1 + 1\times2^0$$
$$= 8+2+1$$
$$= 11$$

$$11100111_2 = 1\times2^7 + 1\times2^6 + 1\times2^5 + 0\times2^4 + 0\times2^3 + 1\times2^2 + 1\times2^1 + 1\times2^0$$
$$= 128+64+32+4+2+1$$
$$= 231$$

Check: $21 \times 11 = 231$

Review Section 1.5

13. (a) Both can be represented as ratios of integers.
$$6.55 = 6\frac{55}{100} = 6\frac{11}{20} = \frac{131}{20}$$

$$6.555... = 6\frac{5}{9} = \frac{59}{9}$$

(b) 3.14 can be represented as a ratio of integers so it is a rational number.

$$3.14 = 3\frac{14}{100} = \frac{314}{100} = \frac{157}{50}$$

π can not be expressed as a ratio of integers so it is irrational.

(c) $-\sqrt{9}$ is the real number -3. $\sqrt{-9}$ is an imaginary number.

14. (a) Natural numbers: $2, \sqrt[3]{125}$

(b) Whole numbers: $0, 2, \sqrt[3]{125}$

(c) Integers: $-7, -\sqrt{16}, 0, 2, \sqrt[3]{125}$

(d) Rational numbers: $-7.6, -7, -5\frac{2}{3}, -\sqrt{16}, 0, 2, 8.33..., 10\frac{3}{5}, \sqrt[3]{125}$

(e) Irrational numbers: $\sqrt[3]{127}$

(f) Real numbers: all numbers except $\sqrt{-16}$

(g) imaginary numbers: $\sqrt{-16}$

(h) Non-integer rational numbers: $-7.6, -5\frac{2}{3}, 8.33..., 10\frac{3}{5}$

15. 400 — composite, abundant
    461 — prime, deficient
    496 — composite, perfect
    512 — composite, deficient

16. By substituting $n = 0$ to $n = 9$, we get 41, 43, 47, 53, 61, 71, 83, 97, 113.
    If n is a multiple of 41, then all three terms of the polynomial will have 41 as a factor.
    Thus, the polynomial will have a factor of 41 and will not result in a prime number.

17. sixth triangular number: 21
    sixth square number: 36
    sixth pentagonal number: 51

CHAPTER 1 TEST

1. ⚲⚲|||

2. MMIII

3. $$=$$
   $$+$$
   $$\equiv$$

   $2003 = 2 \times 1000 + 3$

4. 'βγ

5. · · ↟↟↟↟↟

6. ◄ ◄ ◄ ▼ ▼ ▼   ◄ ◄ ▼ ▼ ▼

$$
\begin{array}{r}
33 \\
60\overline{)2003} \\
\underline{1980} \\
23
\end{array}
\qquad
\begin{array}{r}
23 \\
1\overline{)23} \\
\underline{23} \\
0
\end{array}
$$

7.
$$\begin{array}{r} 5 \\ 360\overline{)2003} \\ \underline{1800} \\ 203 \end{array}\qquad \begin{array}{r} 10 \\ 20\overline{)203} \\ \underline{200} \\ 3 \end{array}\qquad \begin{array}{r} 3 \\ 1\overline{)3} \\ \underline{3} \\ 0 \end{array}$$

8. $11\ 111\ 010\ 011_2$

$$\begin{array}{r} 1 \\ 1024\overline{)2003} \\ \underline{1024} \\ 979 \end{array}\ \begin{array}{r} 1 \\ 512\overline{)979} \\ \underline{512} \\ 467 \end{array}\ \begin{array}{r} 1 \\ 256\overline{)467} \\ \underline{256} \\ 211 \end{array}\ \begin{array}{r} 1 \\ 128\overline{)211} \\ \underline{128} \\ 83 \end{array}\ \begin{array}{r} 1 \\ 64\overline{)83} \\ \underline{64} \\ 19 \end{array}\ \begin{array}{r} 0 \\ 32\overline{)19} \\ \underline{0} \\ 19 \end{array}\ \begin{array}{r} 1 \\ 16\overline{)19} \\ \underline{16} \\ 3 \end{array}\ \begin{array}{r} 0 \\ 8\overline{)3} \\ \underline{0} \\ 3 \end{array}\ \begin{array}{r} 0 \\ 4\overline{)3} \\ \underline{0} \\ 3 \end{array}\ \begin{array}{r} 1 \\ 2\overline{)3} \\ \underline{2} \\ 1 \end{array}\ \begin{array}{r} 1 \\ 1\overline{)1} \\ \underline{1} \\ 0 \end{array}$$

9. $31\ 003_5$

$$\begin{array}{r} 3 \\ 625\overline{)2003} \\ \underline{1875} \\ 128 \end{array}\qquad \begin{array}{r} 1 \\ 125\overline{)128} \\ \underline{125} \\ 3 \end{array}\qquad \begin{array}{r} 0 \\ 25\overline{)3} \\ \underline{0} \\ 3 \end{array}\qquad \begin{array}{r} 0 \\ 5\overline{)3} \\ \underline{0} \\ 3 \end{array}\qquad \begin{array}{r} 3 \\ 1\overline{)3} \\ \underline{3} \\ 0 \end{array}$$

10. $7D3_{16}$

$$\begin{array}{r} 7 \\ 256\overline{)2003} \\ \underline{1792} \\ 211 \end{array}\qquad \begin{array}{r} 13 \\ 16\overline{)211} \\ \underline{208} \\ 3 \end{array}\qquad \begin{array}{r} 3 \\ 1\overline{)3} \\ \underline{3} \\ 0 \end{array}$$

11. $10.666...$

$$10\frac{2}{3} = 10.666...$$

12. XS..

13. $\cap$ ⊚ ⊚
    |||||| ||

$$\frac{2}{3} = \frac{4}{6} = \frac{1}{6} + \frac{3}{6} = \frac{1}{6} + \frac{1}{2}$$

14. ◄ ⟨⟨ ◄ ◄ ◄ ◄

$$\frac{2}{3} = \frac{40}{60}$$

15. ιβ γ' γ'

16. $12{,}231.5_{12}$

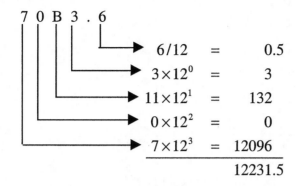

$$6/12 = 0.5$$
$$3 \times 12^0 = 3$$
$$11 \times 12^1 = 132$$
$$0 \times 12^2 = 0$$
$$\underline{7 \times 12^3 = 12096}$$
$$12231.5$$

17. 753_8 , $1EB_{16}$

$$\underbrace{111}_{\substack{4+2+1 \\ 7}}\ \underbrace{101}_{\substack{4+0+1 \\ 5}}\ \underbrace{011}_{\substack{0+2+1 \\ 3}} = 753_8$$

$$\underbrace{1}_{1}\ \underbrace{1110}_{\substack{8+4+2+0 \\ E \\ (14)}}\ \underbrace{1011}_{\substack{8+0+2+1 \\ B \\ (11)}} = 1EB_{16}$$

18. $1\ 010\ 100_2$

$$\begin{array}{r} {}^{1}\ {}^{1}\ {}^{1}\ {}^{1}\ {}^{1}\ \ \\ 1\ 1\ 1\ 1\ 0\ 1 \\ +\quad 1\ 0\ 1\ 1\ 1 \\ \hline 1\ 0\ 1\ 0\ 1\ 0\ 0 \end{array}$$

$$\begin{aligned} 111\ 101_2 &= 1 \times 2^5 + 1 \times 2^4 + 1 \times 2^3 + 1 \times 2^2 + 0 \times 2^1 + 1 \times 2^0 \\ &= 32 + 16 + 8 + 4 + 1 \\ &= 61 \\ 10\ 111_2 &= 1 \times 2^4 + 0 \times 2^3 + 1 \times 2^2 + 1 \times 2^1 + 1 \times 2^0 \\ &= 16 + 4 + 2 + 1 \\ &= 23 \\ 1010100_2 &= 1 \times 2^6 + 0 \times 2^5 + 1 \times 2^4 + 0 \times 2^3 + 1 \times 2^2 + 0 \times 2^1 + 0 \times 2^0 \\ &= 64 + 16 + 4 \\ &= 84 \end{aligned}$$

Check: $61 + 23 = 84$

19. $101\,111\,001_2$

$$
\begin{array}{r}
1\;1\;1\;0\;1 \\
\times \quad 1\;1\;0\;1 \\
\hline
{}^1 1\;{}^1 1\;1\;0\;1 \\
{}^1 0\;0\;0\;0\;0 \\
{}^1 1\;1\;1\;0\;1 \\
{}^1 1\;1\;1\;0\;1 \\
\hline
1\;0\;1\;1\;1\;1\;0\;0\;1
\end{array}
$$

$$
\begin{aligned}
11\,101_2 &= 1\times 2^4 + 1\times 2^3 + 1\times 2^2 + 0\times 2^1 + 1\times 2^0 \\
&= 16+8+4+1 \\
&= 29 \\
1\,101_2 &= 1\times 2^3 + 1\times 2^2 + 0\times 2^1 + 1\times 2^0 \\
&= 8+4+1 \\
&= 13 \\
101\,111\,001_2 &= 1\times 2^8 + 0\times 2^7 + 1\times 2^6 + 1\times 2^5 + 1\times 2^4 + 1\times 2^3 + 0\times 2^2 + 0\times 2^1 + 1\times 2^0 \\
&= 256+64+32+16+8+1 \\
&= 377
\end{aligned}
$$

Check: $29 \times 13 = 377$

20. (a) $\sqrt{-400}$

(b) $-5,\ \sqrt[3]{-27},\ 0,\ 13$

(c) $-\sqrt{40},\ \sqrt[4]{19},\ \pi$

(d) 13

(e) all numbers except those listed in (c) and $\sqrt{-400}$

(f) all numbers except $\sqrt{-400}$

(g) 0, 13

CHAPTER 2 LOGICAL THINKING

Logical thinking is a key to making sound decisions and solving complex problems. Logic is used in every day events such as determining why a car won't start, or planning the route to take when shopping in the city, or filing out a 1040 Tax Form in April. In this chapter, you will examine a few of the many facets of logic in the hope of becoming a better thinker and problem solver. You will begin this journey by looking at some basic ideas about premises and conclusions, proceed to methods of logical argument, analyze truth tables and flow charts, and end with the use logic to help solve puzzles.

SECTION 2.1 LOGIC, STATEMENTS, AND DEFINITIONS

This section discusses the basic components of logic – statements and definitions. These ideas will be the building blocks for all the topics in the chapter.

A **statement** has a clear, definite meaning and a fixed truth value. It cannot be a question, or an opinion, and is either true or false. The types of statements in this section are negations, conditionals, converses, inverses, contrapositives, and biconditionals.

The **negation** of a statement A, $\sim A$, is the statement that has the opposite truth value from A. If A is true, then $\sim A$ is false, and vice versa.

A **conditional** statement is one that has the form *if A then B* and can be symbolized by $A \rightarrow B$.

The **converse** statement of "if A then B" is *if B then A*.

The **inverse** statement of "if A then B" is *if ~A then ~B*.

The **contrapositive** statement of "if A then B" is *if ~B then ~A*. It is important to remember that a statement and its contrapositive always have the same truth values.

A **biconditional** statement is one of the form *A if and only if B*. It is equivalent to the statements "if A then B" and "if B then A".

A **definition** is a statement that names the word being defined, uses terms already understood, and is biconditional.

Explain

1. A statement must have a clear meaning and must be either true or false. It cannot be vague, a question, or a command. Non-statements fail to meet one or more of these conditions.

3. The inverse of the conditional $A \rightarrow B$ is formed by negating A and B, giving
    $\sim A \rightarrow \sim B$.

5. The negation of a statement is formed in several ways. For a simple statement such as "Juan is running," the negation is formed by negating the verb, that is "Juan is not running."

   For statements that involve "all" or "every" the negation is formed by using "some … not." For example, the negation of "All students study biology." is "Some students do not study biology."

   For statements that involve "some … not" the negation is formed by using "all/every." For example, the negation of "Some students do not study biology." is "All student study biology."

   For statements that involve "none" or "no" the negation is formed by using "some." For example, the negation of "No students went to the beach." is "Some students went to the beach."

   For statements that involve "some " the negation is formed by using "none" or "no." For example, the negation of "Some students drink coffee." is "No student drinks coffee."

7. It uses words in the definition that may not be understood.

9. It does not name the term being defined.

11. It is not biconditional.

13. It is not biconditional.

15. The term being defined, "puppy", is named. The words used in the definition should be understood. The statement is biconditional.

17. The term being defined, "quadratic equation", is named. The words used in the definition should be understood. The statement is biconditional.

Apply

19. It is not a statement because it is neither true nor false.

21. It is not a statement because it is neither true nor false.

23. statement

25. It is not a statement because it is neither true nor false.

27. My car is not in the shop.

29. I don't hate sitting around doing nothing.

31. That is not an example of an exponential equation.

33. Some fish cannot live under water.

35. All numbers are prime numbers.

37. No trees are always green.

39. Some of the numbers are positive.

41. Converse: If the phone is in use, then you get a busy signal.
Inverse: If you do not get a busy signal, then the phone is not in use.
Contrapositive: If the phone is not in use, then you do not get a busy signal.

43. Converse: If the point is 16" from the center of the circle, then it is on the circle.
Inverse: If it is not a point on the circle, then it is not 16" from the center of the circle.
Contrapositive: If the point is not 16" from the center of the circle, then it is not on the circle.

45. Converse: If I am listening, then G. H. Mutton is speaking.
Inverse: If G. H. Mutton is not speaking, then I am not listening.
Contrapositive: If I am not listening, then G. H. Mutton is not speaking.

47. Converse: If the figure is not a hexagon, then it has five sides.
Inverse: If the figure does not have five sides, then it is a hexagon.
Contrapositive: If a figure is a hexagon, then it does not have five sides.

Explore

In Problems 49 – 52, and 54 – 59, there are many other correct answers.

49. If x is an integer, then x is a real number.

51. If the product of two real numbers is zero, then at least one of them is zero.

53. not possible

55. If x is a number, then x is a house.

57. If x is a whole number, then x is a prime number.

59. If x is a prime number, then x is a composite number.

61. A dog is a domesticated animal related to the wolf, fox, and jackal.

63. A book is a number of pieces of paper with printing or writing fastened together on one edge.

65. Answers will vary.

67. The contrapositive of the inverse of $A \rightarrow B$ is the converse of $A \rightarrow B$, namely $B \rightarrow A$.

SECTION 2.2 INDUCTIVE AND DEDUCTIVE REASONING

This section discusses two types of reasoning, inductive and deductive.

Inductive reasoning is the type of reasoning where conclusions are formed based on experimentation or experience. This is the type of reasoning that is used to extend existing knowledge into new areas. It does not, however, prove that a conclusion is true. Inductive reasoning suggests that the conclusion is probably true.

Deductive reasoning is the process of reasoning where conclusions are based on accepted premises and the argument follows specific forms. A common form of argument is the **syllogism**. A basic syllogism starts with two premises and draws a conclusion from these premises.

This section discusses three types of syllogisms: hypothetical syllogisms, affirming the antecedent, and denying the consequent. If A, B, and C are statements:

A **hypothetical syllogism** has the following form.

$$A \rightarrow B$$
$$B \rightarrow C$$
$$\therefore A \rightarrow C$$

Affirming the antecedent has the following form.

$$A \rightarrow B$$
$$A$$
$$\therefore B$$

Denying the consequent has the following form.

$$A \rightarrow B$$
$$\sim B$$
$$\therefore \sim A$$

Explain

1. Inductive reasoning is the process in which conclusions are based on experimentation or experience.

3. A syllogism is a form of reasoning in which two statements are made and a conclusion drawn from them.

5. Denying the consequent is a logical argument that has the following form.
$$A \rightarrow B$$
$$\sim B$$
$$\therefore \sim A$$

7. The statement "If it is a normal dog, it has four legs." is true but the inverse "If it is not a normal dog, it does not have four legs." is not always true.

Apply

9. It is not correct. It contains a converse error. This would be correct:
When it is midnight, I am asleep.
It is midnight.
Therefore, I am asleep.

11. It is not correct. It contains an inverse error. This would be correct:
If you are a farmer in Polt County, then you grow corn.
Farmer Ron does not grow corn.
Therefore, Farmer Ron is not a farmer in Polt County.

13. It is a correct hypothetical syllogism.

15. If a whole number greater than 2 is even, then it is divisible by 2.
If a whole number is divisible by 2, then it is not a prime number.
Therefore, if a whole number greater than 2 is even, it is not a prime number.

17. Anyone that treats you with kindness is a nice person.
My teacher, Mrs. Santos, is always very kind to me when I am sick.
Therefore, Mrs. Santos is a nice person.

19. If you are serious about school, you will have less time to watch TV.

21. One of many possible answers:
A: x is a whole number.
B: x is an integer.
C: x is a rational number.
D: x is an irrational number.
E: x equals $\sqrt{2}$.

Explore

23. In a triangle, the longest side is opposite the largest angle and the shortest side is opposite the smallest angle.

25. The new quadrilaterals are parallelograms.

27. The angle formed by the line segments connecting the endpoints of the diameter to a point on a semicircle is a right angle.

29. A possible inductive argument:
    The plane ride is long and cramped.
    The hotels in Honolulu are big and impersonal.
    The beaches in Honolulu are noisy and crowded.
    The restaurants in Honolulu are often expensive.
    Therefore, you should not go to Honolulu, Hawaii.

31. (a)   This is an inductive argument.
    (b)   This is a deductive argument.

33. Some numbers are rational.

35. If the car is full of gasoline, we can see Vernal Falls.

37. By using the contrapositive of the third statement, we get, $P \rightarrow \sim S$.

39. One possible example:
    P = The object is a square.
    Q = The object is a rectangle with four congruent sides.
    R = The object is a triangle.
    S = The object has three sides.
    Conclusion: If an object is a square, then it does not have three sides.

This section introduces the science of symbolic logic and truth tables. **Symbolic logic** is the system of logic that uses symbols to represent statements. A **truth table** is a chart consisting of all possible truth values of the clauses in a statement.

The symbols used in symbolic logic are:

English	Symbol	Name
Not	~	Negation
Therefore	∴	
Implies, If … then …	→	Conditional
Equivalent statements	≡	Logical Equivalency
And	∧	Conjunction
Or	∨	Inclusive Disjunction

The three basic truth tables of symbolic logic are as follows.

Disjunction Truth Table

A	*B*	*A* ∨ *B*
T	T	T
T	F	T
F	T	T
F	F	F

Conjunction Truth Table

A	*B*	*A* ∧ *B*
T	T	T
T	F	F
F	T	F
F	F	F

Conditional Truth Table

A	*B*	*A* → *B*
T	T	T
T	F	F
F	T	T
F	F	T

Truth tables can be used to determine if two statements are logically equivalent and to determine if a syllogism or an argument is correct.

Explain

1. A disjunction is an "or" statement. It is represented by the symbol "∨.

3. A conditional is an "if … then" statement. It is represented by the symbol "→".

5. A truth table is a chart consisting of the possible true and false combinations of the clauses in a statement. Truth tables are used to show that two statements are logically equivalent or to show that an argument is correct.

Apply

7. M = it is midnight, A = asleep
$$M \to A$$
$$A$$
$$\therefore M$$

9. F = farmer in Polt County, C = grows corn
$$F \to C$$
$$\sim C$$
$$\therefore \sim F$$

11. S = scalene, W = two equal sides, H = three equal sides
$$\sim S \to (W \lor H)$$
$$\sim H$$
$$\therefore W$$

13. W = winning golfer, G = good hand-eye coordination, P = positive attitude
$$W \to (G \land P)$$
$$\therefore (G \land P) \to W$$

15. B = voted for Bush, N = voted for Nader, G = voted for Gore
$$(B \lor N) \to \sim G$$

17.

A	C	$A \to C$	$\sim C$	$(A \to C) \lor \sim C$
T	T	T	F	T
T	F	F	T	T
F	T	T	F	T
F	F	T	T	T

The statement is always true.

19.

A	$\sim A$	C	$\sim A \to C$	$\sim C$	$(\sim A \to C) \to \sim C$
T	F	T	T	F	F
T	F	F	T	T	T
F	T	T	T	F	F
F	T	F	F	T	T

The statement is true whenever C is false.

21.

A	B	~B	A ∧ ~B	~(A ∧ ~B)	~A	~A ∨ B
T	T	F	F	**T**	F	**T**
T	F	T	T	**F**	F	**F**
F	T	F	F	**T**	T	**T**
F	F	T	F	**T**	T	**T**

Since both statements have the same truth values, the statements are equivalent.

23.

A	B	A ∨ B	~B	A ∨ ~B	(A ∨ B) ∧ (A ∨ ~B)
T	T	T	F	T	**T**
T	F	T	T	T	**T**
F	T	T	F	F	**F**
F	F	F	T	T	**F**

Since the statements have the same truth values, the statements are equivalent.

25.

P	Q	P → Q	(P → Q) ∧ P	((P → Q) ∧ P) → Q
T	T	T	T	T
T	F	F	F	T
F	T	T	F	T
F	F	T	F	T

Since the last column is always true, the argument is correct.

27.

A	~A	Q	A → Q	~A ∧ (A → Q)	(~A ∧ (A → Q)) → A
T	F	T	T	F	T
T	F	F	F	F	T
F	T	T	T	T	F
F	T	F	T	T	F

Since the last column is not always true, the argument is not correct.

Explore

29. E = Elections become TV popularity contests, G = good looking, smooth talking candidates get elected

$$E \rightarrow G$$
$$\therefore \sim E \rightarrow \sim G$$

E	G	E → G	~E	~G	~E → ~G	(E → G) → (~E → ~G)
T	T	T	F	F	T	T
T	F	F	F	T	T	T
F	T	T	T	F	F	F
F	F	T	T	T	T	T

Since the last column is not always true, the argument is not correct.

31. M = High school graduates have poor math skills
    W = High school graduates have poor writing skills
    J = less able to get a job in the computer industry

$$M \rightarrow J$$
$$W \rightarrow J$$
$$\therefore M \rightarrow W$$

M	W	J	$M\rightarrow J$	$W\rightarrow J$	$(M\rightarrow J)\wedge(W\rightarrow J)$	$M\rightarrow W$	$((M\rightarrow J)\wedge(W\rightarrow J))\rightarrow(M\rightarrow W)$
T	T	T	T	T	T	T	T
T	T	F	F	F	F	T	T
T	F	T	T	T	T	F	F
T	F	F	F	T	F	F	T
F	T	T	T	T	T	T	T
F	T	F	T	F	F	T	T
F	F	T	T	T	T	T	T
F	F	F	T	T	T	T	T

Since the last column is not always true, the argument is not correct.

33. Constructing a truth table shows that the argument is incorrect if R is true, N is false and S is false.

Flowcharts describe the logical path that is followed when a decision is made. They are also used to display organizational structures.

There are four basic symbols used in flowcharts.

The **start/stop symbol** is a circle or oval. It indicates the beginning of end of a line of logic.

The **statement symbol** is a rectangle. It indicates either an action, person or result.

The **decision symbol** is a diamond shaped object known as a rhombus. The decision symbol indicates a question in the flow chart.

The final symbol, the **flow line**, is a ray that connects the other symbols and describes the path that will be followed as you progress through the flowchart.

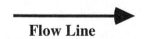

Explain

1. Flowcharts describe the logical path that is followed when a decision is made. They are also used to display organizational structures.

3. The **decision symbol** is a diamond shaped object known as a rhombus. The decision symbol indicates a question in the flowchart. The symbol use:

5. An organizational chart displays the management or command structure of a corporation, school, or government bureau.

7. "A flowchart is a picture of logic in use" because it gives a visual representation of the logical pattern in a decision making process.

Apply

9. This flowchart describes the process which helps decide why a lawn mower will not start. The choices are limited to the lawn mower being out of gas, the spark plug wire being disconnected, the blade being engaged, or some unknown reason.

11. (a) Stand.
    (b) Stand if the dealer shows a card $2 \leq x < 7$, otherwise draw another card and then reevaluate your hand.
    (c) Draw another card and then reevaluate your hand.
    (d) Draw another card and then reevaluate your hand.
    (e) Stand.
    (f) To show how to win the game, add the following to the existing flowchart.

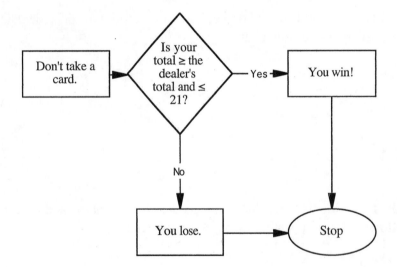

Explore

13.

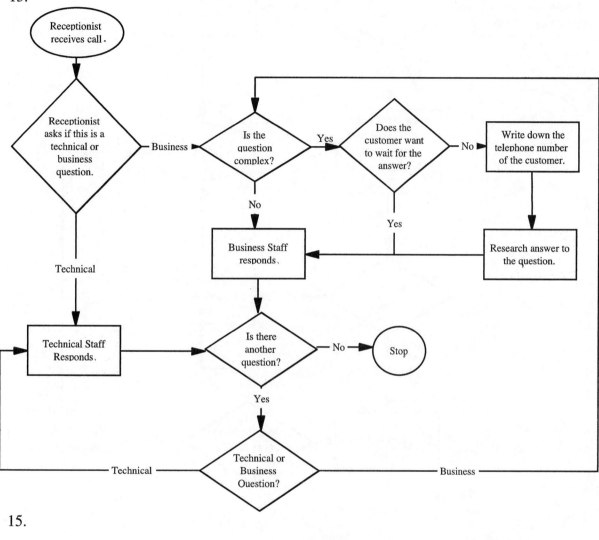

15.

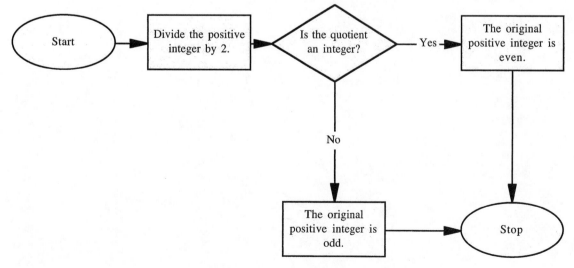

17.

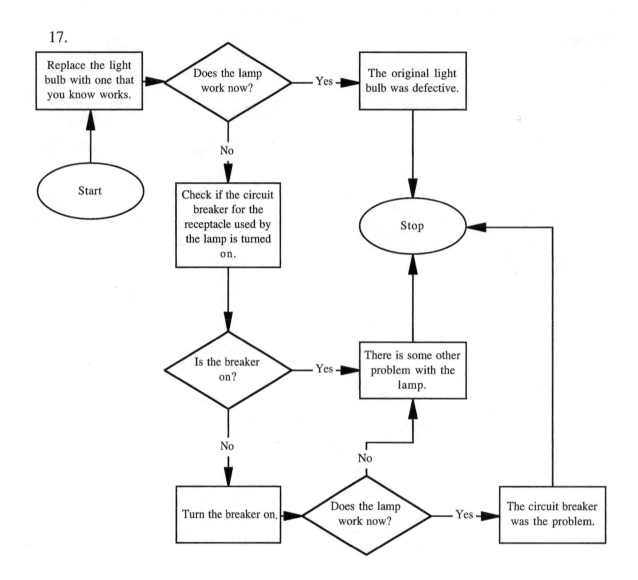

19.

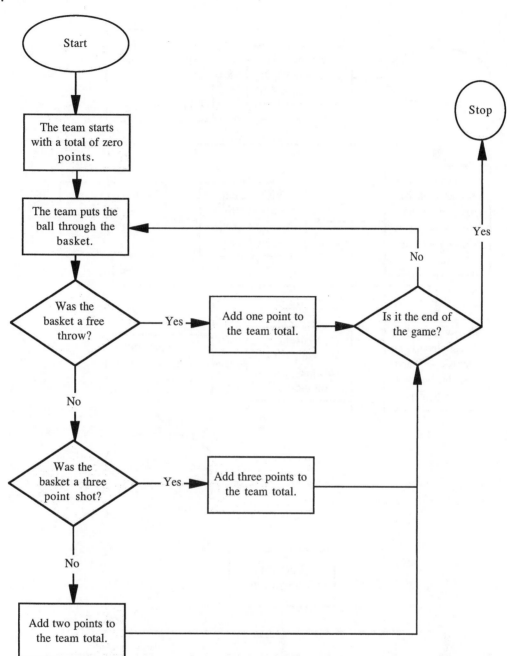

21.

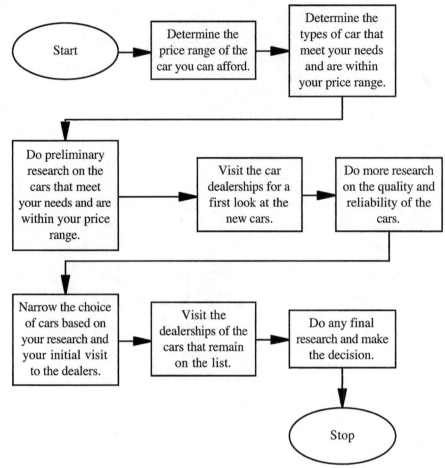

23. Answers will vary.

25.

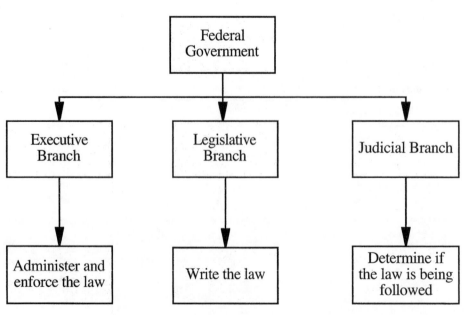

In this section, you are introduced to various types of puzzles in which the use of logic is essential in arriving at correct solutions. Magic Triangles, Magic Squares, Alphametic Puzzles, Cross Number Puzzles and others are presented. Hints on the logic used in solving these puzzles are given. These suggestions along with "educated" trial and error techniques will be used in solving the puzzles. You will be challenged to have fun and solve some of them on your own. The answers to many of the puzzles are not unique so their solutions should lead to interesting discussions.

Explain

1. A magic square is a square array of numbers, 1, 2, 3, 4, 5, ..., that has the same sum horizontally, vertically, and diagonally.

3. Most of these puzzles have many possible solutions so using just trial and error might take you a long time to solve the puzzle.

5. E = 0, since C − E = C in the ones column.
    A = 1, since there is no digit in the hundreds column of the answer, borrowing must have been used in the 10's column.

7. E = 0, since the only number that produces the same digit in all places when multiplying is zero. Further, since E + C = C and E + B = B, E = 0. D = 1, since ABC × D = ABC.

Apply

9. Sum of numbers equals 18, 20, 22, and 24.

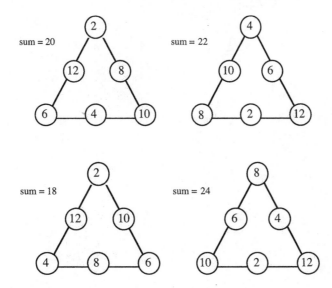

11. Sum of numbers in any row, column, or diagonal equals 27. Replace the number in each square of the 3 × 3 magic square with twice the number decreased by 1. If n = number in the magic square, replace it with $2n - 1$.

15	5	7
1	9	17
11	13	3

13.

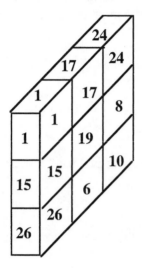 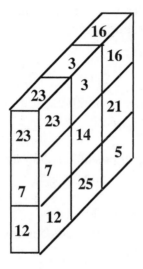 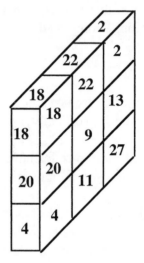

15. One of the possible solutions:

$$
\begin{array}{ccc}
 A & B & C \\
 - & D & E \\
\hline
 & G & C \\
\end{array}
\quad \Rightarrow \quad
\begin{array}{ccc}
 1 & 3 & 2 \\
 - & 4 & 0 \\
\hline
 & 9 & 2 \\
\end{array}
$$

$$
\begin{array}{r}
GA \\
ABC\overline{)ADEC} \\
ABC \\
\hline
FEC \\
FEC \\
\hline
\end{array}
\quad \Rightarrow \quad
\begin{array}{r}
12 \\
230\overline{)2760} \\
230 \\
\hline
460 \\
460 \\
\hline
\end{array}
$$

Explore

17.   Possible sum of legs are 25, 23, 27, 24, and 26.

		9		
		7		
2	6	5	4	8
		3		
		1		

		9		
		7		
8	6	1	5	3
		4		
		2		

		8		
		3		
2	5	9	4	7
		6		
		1		

		5		
		7		
2	4	3	6	9
		1		
		8		

		6		
		8		
1	5	7	4	9
		2		
		3		

19. The sum of the numbers on each line segment is 8.

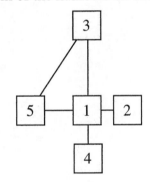

21. One possible solution:

```
    G  O  L  F          7  4  3  0
  + R  A  I  N   ⇒   +  6  5  9  1
  ─────────────      ──────────────
    N  O  F  U  N     1  4  0  2  1
```

23. One possible solution:

```
    F  I  V  E          3  4  7  5
  − F  O  U  R   ⇒   −  3  2  6  0
  ─────────────      ──────────────
       O  N  E              2  1  5
```

25. One possible solution:

```
    Y  O  U             5  0  4
  ×       R     ⇒   ×        2
  ───────────       ──────────
    C  O  O  L        1  0  0  8
```

27. One possible solution:

$$
\begin{array}{ccccc}
 & \text{B} & \text{I} & \text{G} \\
\times & \text{M} & \text{A} & \text{C} \\
\hline
 & \text{B} & \text{I} & \text{G} \\
 \text{A} & \text{A} & \text{A} & \\
\text{Y} \ \text{E} & \text{S} & \text{A} & \\
\hline
\text{Y} \ \text{E} & \text{S} & \text{B} & \text{I} \ \text{G}
\end{array}
\qquad \Rightarrow \qquad
\begin{array}{cccccc}
 & & 4 & 7 & 6 \\
 & \times & 5 & 0 & 1 \\
\hline
 & & 4 & 7 & 6 \\
 & 0 & 0 & 0 & \\
2 & 3 & 8 & 0 & \\
\hline
2 & 3 & 8 & 4 & 7 \ 6
\end{array}
$$

29. One possible solution:

$$
\begin{array}{cccc}
 & \text{S} & \text{E} & \text{N} & \text{D} \\
+ & \text{M} & \text{O} & \text{R} & \text{E} \\
\hline
\text{M} & \text{O} & \text{N} & \text{E} & \text{Y}
\end{array}
\quad \Rightarrow \quad
\begin{array}{cccc}
 & 9 & 5 & 6 & 7 \\
+ & 1 & 0 & 8 & 6 \\
\hline
1 & 0 & 6 & 5 & 3
\end{array}
$$

31.

36	÷	6	×	1	**6**
÷		−		+	
9	−	3	×	5	**−6**
−		×		−	
8	−	4	×	2	**0**
−4		**−6**		**4**	

1 2 3 4 5 6 8 9

33.

5	÷	$\sqrt{9}$	×	6	**10**
+		+		−	
1	−	$\sqrt{4}$	−	7	**−8**
−		×		×	
8	+	3	+	2	**13**
−2		**9**		**−8**	

1 2 3 4 5 6 7 8 9

Review Section 2.1

1. Some of my relatives do not live in Argentina.

2. If you fixed it, it was broken.

3. Math is the study of the relationship between quantities using numbers and symbols.

4. A statement in logic can not be true or false at the same time. Questions, commands, and vague sentences are not considered statements.

Review Section 2.2

5. The argument has the converse error. The correct argument is as follows.

    > If there is a drought, you don't water the lawn.
    > You water the lawn.
    > Therefore, there is no drought.

6. By substituting $n = 0, 1, 2, 3, 4, 5, 6, 7, 8, 9, \ldots$, we get 41, 43, 47, 53, 61, 71, 83, 97, 113, which are all prime numbers. You may, therefore, make the conclusion that this formula always produces a prime number. However, when $n = 41$ the answer is 1681 which is a composite number (41×41).

7. Inductive: I had a friend who drank too many beers and got in an accident. He wrecked his car and nearly killed himself. This might happen to you, so you should not drive.

    Deductive: If you drink excessively, your reactions and coordination is impaired.
    If your reactions and coordination is impaired, you may get in an accident.
    If you get in an accident, you could hurt yourself, your car, and others.
    If you could hurt yourself, your car, and others, you should not drive.

    Therefore, if you drink excessively, you should not drive.

8. Rearranging the statements and using the contrapositive for the first given statement, gives:

    > If there are no reflections in the scene, you don't use a polarizing filter.
    > If you don't use a polarizing filter, you are not photographing water.
    > Therefore, if there are no reflections in the scene, you are not photographing water.

9. D = dog is a Brittany
    H = hunt birds

    $D \rightarrow H$

    D

    $\therefore H$

10. The statement is true if B is true.

A	B	$\sim A$	$\sim B$	$\sim A \rightarrow \sim B$	B	$(\sim A \rightarrow \sim B) \rightarrow B$
T	T	F	F	T	T	T
T	F	F	T	T	F	F
F	T	T	F	F	T	T
F	F	T	T	T	F	F

11. Yes, the statements are equivalent.

A	B	$A \wedge B$	$\sim (A \wedge B)$	$\sim A$	$\sim B$	$\sim A \vee \sim B$
T	T	T	**F**	F	F	**F**
T	F	F	**T**	F	T	**T**
F	T	F	**T**	T	F	**T**
F	F	F	**T**	T	T	**T**

12. The argument is not correct. The last column does not contain all true values.
    Let R = election reforms
    B = candidates buy elections

R	B	$\sim R$	$\sim R \rightarrow B$	$\sim B$	$R \rightarrow \sim B$	$(\sim R \rightarrow B) \rightarrow (R \rightarrow \sim B)$
T	T	F	T	F	F	F
T	F	F	T	T	T	T
F	T	T	T	F	T	T
F	F	T	F	T	T	T

Review Section 2.4

13. If your printer is not working you need to check and correct possible problems: Is the computer and printer turned on? Are the computer and printer plugged in? Are the printer cable connected to the computer? Does the printer have paper? If all these are checked and corrected, get some help from an expert. The flow chart simply organizes the process and gives a logical order to solving the problem.

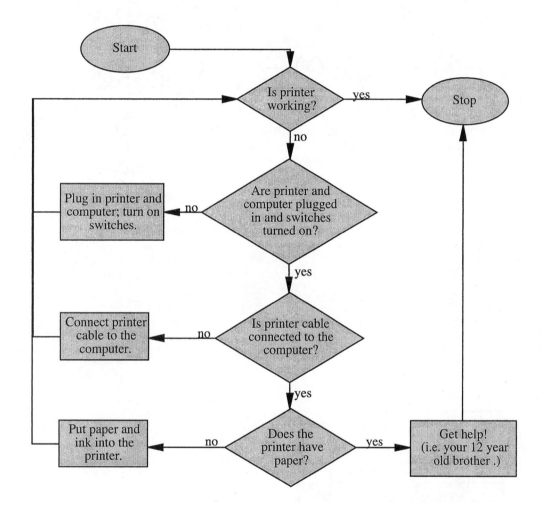

14.

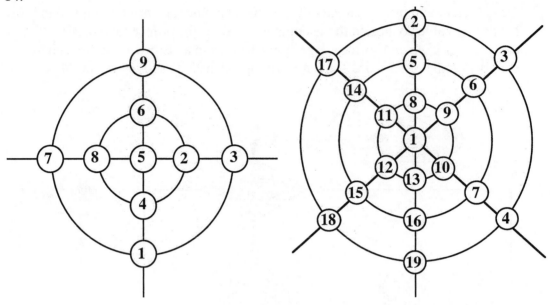

15.

$$\begin{array}{cccc} & T & E & E \\ & H & E & E \\ + & H & E & E \\ \hline J & O & K & E \end{array} \qquad \Rightarrow \qquad \begin{array}{cccc} & 4 & 5 & 5 \\ & 7 & 5 & 5 \\ + & 7 & 5 & 5 \\ \hline 1 & 9 & 6 & 5 \end{array}$$

16.

$$\begin{array}{ccc} F & U & N \\ \times & & N \\ \hline S & U & N \end{array} \qquad \Rightarrow \qquad \begin{array}{ccc} 1 & 2 & 5 \\ \times & & 5 \\ \hline 6 & 2 & 5 \end{array}$$

17.

36	÷	4	×	9	**81**
÷		+		×	
2	+	7	−	8	**1**
÷		−		÷	
6	×	1	÷	3	**2**
3		**10**		**24**	

1. (a) Hypothesis: "If it is after midnight." Conclusion: "I am in bed."
   (b) If I am in bed, it is after midnight.
   (c) If it is not after midnight, then I am not in bed.
   (d) If I am not in bed, then it is not after midnight.

2. Three components of a definition and the additional properties needed for a good definition can be found in Section 2.1 under definitions.
   (a) Not biconditional. Perpendicular lines are lines that intersect at right angles.
   (b) The statement uses the same root word in its definition and thus, it uses a word in the definition that may not be already understood.
   Microscopic: that which can only be seen through a high powered magnifying device.
   (c) Not biconditional.
   Natural numbers are whole numbers that are greater than zero.

3. One of many possible answers: If x is a natural number, then it is a real number.

4. The difference between inductive and deductive arguments is explained in Section 2.2. There are many possible arguments. Remember that an inductive argument is based on observing results, analyzing experiences, citing authorities, or presenting statistics. A deductive argument follows from accepted facts, assumptions, rules, or laws.

5. (a) Using the rule for contrapositives: $Z \rightarrow \sim Y$ is equivalent to $Y \rightarrow \sim Z$.

   Thus, the argument is as follows.
   $$X \rightarrow Y$$
   $$Y \rightarrow \sim Z$$
   $$\sim Z \rightarrow P$$
   $$\therefore \ X \rightarrow P$$

   (b) Using the rule for contrapositives: "If the geometry is Euclidean, then parallel lines exist." is equivalent to "If there are no parallel lines, then the geometry is non-Euclidean."

   Thus the argument is as follows.

   If the geometry is Riemannian, then there are no parallel lines.
   If there are no parallel lines, then the geometry is non-Euclidean.
   If the geometry is non-Euclidean, then at least one of Euclid's postulates is changed.
   $\therefore$ If the geometry is Riemannian, then at least one of Euclid's postulates is changed.

6. (a) Let H = headache, G = grumpy, S = silent, $H \rightarrow (G \vee S)$
   (b) Let G = good weather, B = play baseball, P = have a picnic, $\sim G \rightarrow (\sim B \wedge \sim P)$
   (c) Let M = study math, G = good job, H = harder to advance, $(\sim M \wedge G) \rightarrow H$

7.

A	B	$A \vee B$	$\sim B$	$A \vee \sim B$	$(A \vee B) \wedge (A \vee \sim B)$
T	T	T	F	T	**T**
T	F	T	T	T	**T**
F	T	T	F	F	**F**
F	F	F	T	T	**F**

Since the two statements have the same truth values, the statements are equivalent.

8. Let P = parrot, W = cracks walnuts, A = my animal
   The argument is $(P \rightarrow W) \wedge (A \rightarrow \sim W)$
   $$A \rightarrow \sim P$$

P	W	A	$P \rightarrow W$	$\sim W$	$A \rightarrow \sim W$	$(P \rightarrow W) \wedge (A \rightarrow \sim W)$	$\sim P$	$A \rightarrow \sim P$	$((P \rightarrow W) \wedge (A \rightarrow \sim W)) \rightarrow (A \rightarrow \sim P)$
T	T	T	T	F	F	F	F	F	T
T	T	F	T	F	T	T	F	T	T
T	F	T	F	T	T	F	F	F	T
T	F	F	F	T	T	F	F	T	T
F	T	T	T	F	F	F	T	T	T
F	T	F	T	F	T	T	T	T	T
F	F	T	T	T	T	T	T	T	T
F	F	F	T	T	T	T	T	T	T

Since the last column is always true, the argument is correct.

9.

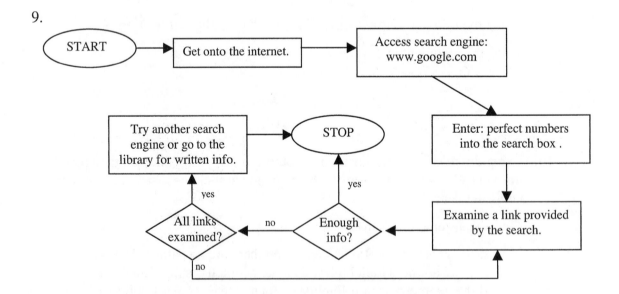

10.

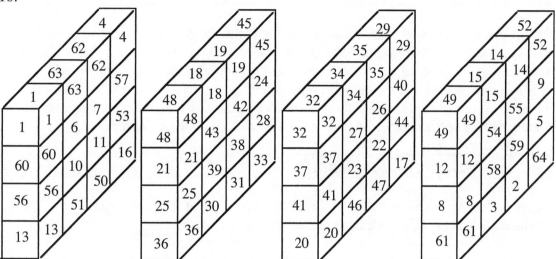

11.

$$
\begin{array}{cccc}
 & Y & E & A \\
\times & & & 4 \\
\hline
M & A & T & H
\end{array}
\Rightarrow
\begin{array}{cccc}
 & 9 & 6 & 8 \\
\times & & & 4 \\
\hline
3 & 8 & 7 & 2
\end{array}
$$

12.

72	÷	8	×	1	9
÷		−		×	
9	−	3	×	4	−3
−		−		+	
2	×	5	−	0	10
6		0		4	

0 1 2 3 4 5 8 9

CHAPTER 3 SETS AND COUNTING

In this chapter, you will learn about sets and how the abstract mathematical concept of a set is applied to real-world problems. In particular, you will become familiar with the terminology and operations of sets such as unions and intersections. You will see how Venn diagrams give a graphical representation of sets. Finally, you will see how the number of items in a set is directly related to counting things like the number of possible lottery tickets or the number of car license plates in the state of California.

SECTION 3.1 SETS, FINITE AND INFINITE

A **set** is a collection of items. An **element** of a set is one of the items contained in the set. To indicate that an item x is an element of a set A, we write $x \in A$.

The number of elements in the set is called the **cardinal number** of a set and is written $n(S)$. If the cardinal number of a set is a whole number, the set is a **finite set**. If a set has an unlimited number of elements, it is an **infinite set**.

Set B is a **subset** of set A if every element of B is also an element of A. To indicate that B is a subset of A, we write $B \subseteq A$. The set B is called a **proper subset** of A if $n(B) < n(A)$. A proper subset is denoted by $B \subset A$.

Two sets are said to be **equal** if the have exactly the same elements. Two sets are said to be **equivalent** if they have the same *number* of elements, that is $n(A) = n(B)$.

Two sets A and B are said to be equivalent if you can set up some relationship between them so that for every element of A, there is exactly one element in B and for each element in B, there is exactly one element in A. This type of relationship is called a **one-to-one** relationship.

Explain

1. A set is a collection of items.

3. When using the descriptive method, the set is given by describing the contents of the set rather than listing each element, i.e. "whole numbers less than 6" rather than {0, 1, 2, 3, 4, 5}.

5. Set A is a subset of set B if every element of A is also an element of B.

7. Two finite sets are said to be equivalent if they have the same *number* of elements, that is $n(A) = n(B)$.

9. A set is countably infinite if it can be put into a one-to-one relationship with the natural numbers.

Apply

11. (a) {*x*|*x* is an even whole number ≤ 10}
    (b) {0, 2, 4, 6, 8, 10}

13. The symbol is being used correctly and the statement is true.

15. The symbol is being used correctly but the statement is false. One possible true statement is $T \not\subset F$.

17. An incorrect symbol is being used. A true correct statement is $5 \in I$.

19. The symbol is being used correctly but the statement is false. Use $5 \in I$.

21. The symbol is being used correctly and the statement is true.

23. The symbol is being used incorrectly since both *F* and *I* are sets. The proper symbol to use is the subset symbol. One possible correct statement is $F \subset I$.

25. Let $A = \{1,2,3,4,5,...\}$ and $B = \{5,10,15,20,25,...\}$. We can set up the relationship
    $$A: \quad 1 \quad 2 \quad 3 \quad 4 \quad 5 \quad ... \quad k$$
    $$\updownarrow \quad \updownarrow \quad \updownarrow \quad \updownarrow \quad \updownarrow \qquad \updownarrow .$$
    $$B: \quad 5 \quad 10 \quad 15 \quad 20 \quad 25 \quad ... \quad 5k$$
    Thus for each natural number in *A*, there is exactly one natural number in *B*, and for each $k \in A$, $5k \in B$.

27. Let $A = \{1,2,3,4,5,...\}$ and $B = \{1,4,9,16,25,36,...\}$. We can set up the relationship
    $$A: \quad 1 \quad 2 \quad 3 \quad 4 \quad 5 \quad ... \quad k$$
    $$\updownarrow \quad \updownarrow \quad \updownarrow \quad \updownarrow \quad \updownarrow \qquad \updownarrow .$$
    $$B: \quad 1 \quad 4 \quad 9 \quad 16 \quad 25 \quad ... \quad k^2$$
    Thus for each natural number in *A*, there is exactly one natural number in *B*, and for each $k \in A$, $k^2 \in B$.

Explore

29. Looking in an atlas, we can find that the set of states that border Indiana is
    A = {Michigan, Ohio, Kentucky, Illinois}.

31. N = {} since there are no countries that border South Africa that are also north of the equator.

33. The presidents in office (1940 - 2003) are P = {Roosevelt, Truman, Eisenhower, Kennedy, Johnson, Nixon, Ford, Carter, Reagan, G. H. Bush, Clinton, G. W. Bush} so $n(P)$ = 12.

35. As of 2003, $n(B) = 30$

37. We show that the sets are equivalent by showing that they can be put in one-to-one correspondence with the natural numbers.

$$N: \quad 1 \quad 2 \quad 3 \quad 4 \quad ... \quad k$$
$$\updownarrow \quad \updownarrow \quad \updownarrow \quad \updownarrow \qquad \updownarrow$$
$$R: \quad 3 \quad 6 \quad 9 \quad 12 \quad ... \quad 3k$$
$$\updownarrow \quad \updownarrow \quad \updownarrow \quad \updownarrow \qquad \updownarrow$$
$$S: \quad 1 \quad 3 \quad 9 \quad 27 \qquad 3^{k-1}$$

SECTION 3.2   SET OPERATIONS AND VENN DIAGRAMS

The **complement** of a set A is the set of items in a universal set that are not contained in the set A. The complement is denoted as $\overline{A}$ and is read *not A"* or "A bar".

The **union** of two sets A and B is the set containing all the elements that are in at least one of the two sets A and B. The notation for the union of A and B is $A \cup B$.

The **intersection** of two sets A and B is the set containing all the elements that are in both of the two sets A and B. The notation for the intersection of A and B is $A \cap B$.
When two sets have no intersection, they are called **disjoint sets.**

A **Venn diagram** is a picture which uses geometric shapes to represent sets.

Cardinality of a Union is the number of elements in the union of two sets A and B.
$$n(A \cup B) = n(A) + n(B) - n(A \cap B)$$

Explain

1. The intersection of two sets is the set that contains the items that are in both of the original sets.

3. Two sets are disjoint if they have no intersection.

5. Both $n(A)$ and $n(B)$ include the elements of $A \cap B$ so those elements are counted twice. To eliminate this duplicate counting $n(A \cap B)$ is subtracted once.

7. $n(A \cap B) = n(B)$, all the elements of B must be contained inside of A.

Apply

9. (a) {apples, bananas, peaches, tomatoes, beans, peas, sprouts}
    (b) {tomatoes}

11. $n(A \cup B) = n(A) + n(B) - n(A \cap B) = 25 + 30 - 7 = 48$

13. Using the union formula, we get the following.
$$n(A \cup B) = n(A) + n(B) - n(A \cap B)$$
$$40 = 25 + 30 - n(A \cap B)$$
$$n(A \cap B) = 25 + 30 - 40$$
$$n(A \cap B) = 15$$

15. Using the union formula we get the following.
$$n(A \cup B) = n(A) + n(B) - n(A \cap B)$$
$$60 = 25 + 30 - n(A \cap B)$$
$$n(A \cap B) = 25 + 30 - 60$$
$$n(A \cap B) = -5$$
Since the number of elements in a set cannot be negative, the stated conditions are not possible.

17. $n(A) = 51 + 25 = 76$

19. $n(A \cup B) = 51 + 25 + 17 = 93$

21. $n(A \cap \overline{B}) = 51$

23. $n(\overline{A} \cap \overline{B}) = 7$

Explore

25.

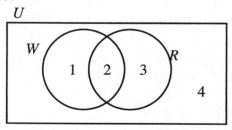

1 = {all women without red hair}
2 = {all women with red hair}
3 = {all people who have red hair but are not women}
4 = {all people who do not have red hair and are not women}

27.

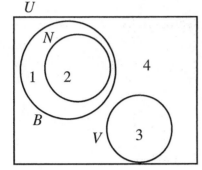

1 = {all books in the library that are not novels}
2 = {all books in the library that are novels}
3 = {all videos in the library}
4 = {all items in the library that are not books and not videos}

29. $x = 300 - 72 - 117 - 43 = 68$

31. $A \cap B$ is the empty set.

33. 125 people

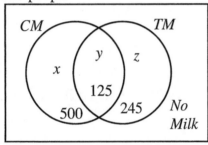

$x + y + z = 1000 - 130 = 870$
$x + y = 625$
$y + z = 370$

Subtracting the first two equations gives $z = 245$. Substituting this value into the third equation gives $y = 125$.

SECTION 3.3 APPLICATIONS OF SETS

In this section, you will apply the concepts of sets to answering questions about real-world data. The two primary techniques will be to read information from tables and the use of Venn diagrams. There is no additional notation or formulas presented in this section.

Explain

1. The notation $T \cup F$ represents the actual set of items contained in the union of two sets. The notation $n(T \cup F)$ represents the number of items in the union.

3. The value a can be found by adding 100 and 85.
    The values 65 and b must add to 130. Therefore, b can be found by subtracting 65 from 130.

5.

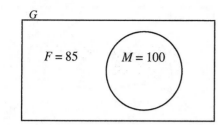

Apply

7. There are a total of 255 doctors who accept children as patients and $40 + 135 + 50 = 225$ of them are either perform surgery or are in general practice. Therefore, $255 - 225 = 30$ only have children as patients.

9. $n(A) = 62$

11. $M \cup A$ is the set of males or those working in farming, forestry or fishing.
$$n(M \cup A) = n(M) + n(A) - n(M \cap A)$$
$$= 1200 + 62 - 47 = 1215$$

13. $M \cup F$ is the set of males or females.
$$n(M \cup F) = n(M) + n(F) - n(M \cap F)$$
$$= 1200 + 1300 - 0 = 2500$$

15. $\overline{M} \cap \overline{A}$ is the set of females who are not working in farming, forestry or fishing.
$$n(\overline{M} \cap \overline{A}) = 1300 - 15 = 1285$$

17. $n(A \cap B \cap C) = 53$

19. $n(\overline{A} \cup B \cup \overline{C}) = 31 + 43 + 53 + 12 + 71 + 17 + 40 = 267$

21. $n(A \cup C) = 17 + 43 + 53 + 58 + 12 + 71 = 254$

23. (a) $A \cap B \cap C = IV$
    (b) $A \cap B \cap \overline{C} = II$
    (c) $A \cap \overline{B} \cap C = III$
    (d) $\overline{A} \cap B \cap C = V$
    (e) $A \cap \overline{B} \cap \overline{C} = I$
    (f) $\overline{A} \cap \overline{B} \cap C = VII$
    (g) $\overline{A} \cap B \cap \overline{C} = VI$
    (h) $\overline{A} \cap \overline{B} \cap \overline{C} = VIII$

Explore

25. (a) $x + 43 + 75 + 50 = 242, x = 74$
     (b) $y + 43 + 75 + 92 = 278, y = 68$
     (c) $z + 50 + 75 + 92 = 298, z = 81$
     (d) $n + 75 + 43 + 50 + 92 + 74 + 68 + 81 = 500, n = 17$

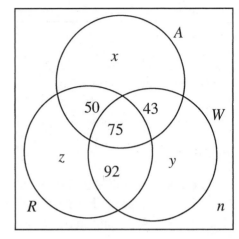

27. For both parts (a) and (b), the shaded region is outside of both A and B.

29. One possible version of the formula is
$$n(A \cup B \cup C) = n(A) + n(B) + n(C) - n(A \cap B) - n(A \cap C) - n(B \cap C) + n(A \cap B \cap C).$$

SECTION 3.4 INTRODUCTION TO COUNTING

This section examines the methods used in determining the number of ways an event can occur. The most straightforward method to do this is to simply list all possible outcomes and count them. When listing the outcomes, it is important to keep in mind if the counting is done with items being replaced or without being replaced. If an item is chosen from a set **with replacement**, the item can be chosen repeatedly from the set. For example, if a coin lands heads on the first toss, it can also lands heads on any future toss. If an item is chosen from a set **without replacement**, the item cannot be chosen from the set repeatedly. For example, if the ace of diamonds is drawn from a standard deck of cards and not returned to the deck, the ace of diamonds cannot be drawn from the deck a second time.

When counting items that are chosen with replacement, the **Basic Counting Law** is used. The Basic Counting Law says that if there are n choices for the first item and m choices for the second item, there are $n \times m$ ways in which to select the two items. For example, if there are three salads and two types of bread, there are $3 \times 2 = 6$ ways to pick one salad and one type of bread. If items are chosen without replacement, permutations and combinations can be used. Permutations are used when items are chosen without replacement and order matters. Combinations are used when items are chosen without replacement and order does *not* matter. The chart that follows will help determine how to approach a given counting problem.

Summary of Counting Principles

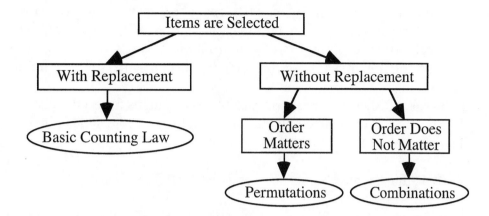

Explain

1. Counting by listing involves writing out all possible outcomes of an event.

3. When an item is chosen from a set with replacement, the item can be chosen repeatedly from the set.

5. The Basic Counting Law says that if there are n choices for the first item and m choices for the second item, there are $n \times m$ ways in which to select the two items.

7. Permutations are used when items are chosen without replacement and order matters.

Apply

9. Permutations or Basic Counting Law

11. Basic Counting Law

13. Combinations

15. Basic Counting Law

17. Permutations or Basic Counting Law

19. Permutations or Basic Counting Law

21. Combinations

Explore

23. The number of ways is $2 \times 2 \times 2 \times 2 = 16$. The possibilities are:
    (on, on, on, on); (on, on, on, off), (on, on, off, on); (on, off, on, on);
    (off, on, on, on); (on, on, off, off); (on, off, on, off); (off, on, off, on);
    (off, off, on, on); (on, off, off, on); (off, on, on, off); (off, off, off, on);
    (off, off, on, off); (off, on, off, off); (on, off, off, off); (off, off, off, off).

25. $10^9 = 1,000,000,000$; Since there are 10 possible digits for each of the nine digits of the Social Security number, by the basic counting law, the number of Social Security numbers is $10 \times 10 \times 10 \times 10 \times 10 \times 10 \times 10 \times 10 \times 10 \times 10 = 10^9$.

27. 3; By listing the possibilities, we get (BBG), (BGB), (GBB).

29. 210; There are 7 candidates in which the order of finish for the top three determines the position held. Thus, the number of ways this can be done is $P_{7,3} = 210$.

31. $2^{10} = 1024$; Since each of the 10 questions has two possible answers, by the basic counting law, the number of ways to answer the questions is
    $2 \times 2 \times 2 \times 2 \times 2 \times 2 \times 2 \times 2 \times 2 \times 2 = 2^{10}$.

33. 2,598,960; Since we are dealt 5 cards from a deck of 52 cards and the order that the cards are received does not matter, there are $C_{52,5} = 2,598,960$ possible hands.

35. (a) 1024; There are four 5's, four 6's, four 7's, four 8's and four 9's in the deck. Using the basic counting law, there are $4 \times 4 \times 4 \times 4 \times 4 = 1024$ ways to get a straight starting with a 5 and ending with a 9.
    (b) 10,240; The highest card in a straight may be any one of ten cards (A, K, Q, J, 10, 9, 8, 7, 6, 5). Since there are 1024 ways to get a straight with a given high card, there are $10 \times 1024 = 10,240$ ways to get any straight.

37. 347,373,600; Since there are 32 cards numbered 2 through 9 and a yarborough consists of 13 of these cards in any order, the number of yarboroughs is
    $C_{32,13} = 347,373,600$.

39. 190; Since the dog must pick two out of 20 items in any order, the number of ways is $C_{20,2} = 190$.

41. 30; Since you can attend two of four poetry readings and one of five other events, the number of ways this can be done is $C_{4,2} \times C_{5,1} = 30$.

43. 175,760,000; By the basic counting law, the number of license plates is
    $10 \times 26 \times 26 \times 26 \times 10 \times 10 \times 10 = 175,760,000$.

45. 180; Since a handshake requires two people, we want the number of combinations of 2 people taken from the 20 people, that is, $C_{20,2}$. However, we need to

exclude 10 of these handshakes since a person does not shake hands with their spouse. Therefore, the number of handshakes is $C_{20,2} - 10 = 180$.

47. 38,760; Out of the 20 winning numbers, you are picking 6. Since the order in which the numbers are selected does not matter, the number of possible tickets is $C_{20,6} = 38,760$.

CHAPTER 3 REVIEW

Review Section 3.1

1. (a) $\{x|x$ is a whole number less than 20 that is divisible by 3$\}$
   (b) $\{0, 3, 6, 9, 12, 15, 18\}$

2. Equivalent sets are sets that contain the same number of elements or sets that can be put in a one-to-one relationship. Equal sets are sets that have the same elements in the sets.

3. The cardinality of a set is the number of elements in a set.

4. A countably infinite set is one that can be put into a one-to-one relationship with the natural numbers. For the set $\{12, 23, 34, 45, \ldots\}$, we have the following.

$$
\begin{array}{cccccccc}
1 & 2 & 3 & 4 & 5 & \ldots & k \\
\updownarrow & \updownarrow & \updownarrow & \updownarrow & \updownarrow & & \updownarrow \\
12 & 23 & 34 & 45 & 56 & \ldots & 1+11k
\end{array}
$$

Review Section 3.2

5. Region a is the set of dogs who like to swim but do not have black hair. Region b is the set of dogs who like to swim and have black hair. Region c is the set of dogs who don't like to swim and have black hair. Region d is the set of dogs that do not like to swim and do not have black hair.

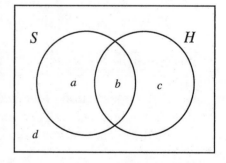

6. $T \cap S$ is the set of objects that are in both T and S. Therefore the set of the first four elements of $T \cap S = \{12, 24, 36, 48\}$.

7. $n(A \cup B) = 51 + 25 + 17 = 93$

8. $n\left(A\cap\overline{B}\right)=51$

Review Section 3.3

9. $G\cap\overline{M}$ is the set of girls who do not prefer movies. $n\left(G\cap\overline{M}\right)=5$

10. $B\cup\overline{V}$ is the set of either boys or people who do not prefer video games.
$n\left(B\cup\overline{V}\right)=6+3+10=19$

11. $n\left(A\cap B\cap\overline{C}\right)=27$. This is the set of items in A and B but not in C.

12. $n\left(\overline{A}\cap\overline{B}\right)=29+23=52$. This is the set of items not in A and not in B.

Review Section 3.4

13. (a) The Basic Counting Law is used when the counting problem involves replacement.
    (b) Permutations are used when the items are not replaced and order matters.
    (c) Combinations are used when the items are not replaced and order does not matter.

14. The number of lunches is $3\times6\times4=72$.

15. There are a total of 20 numbers from which we pick 2, so
number of ways $= C_{20,2}=190$.

16. There are 6 ways to get a sum of 5 on three dice: (1, 1, 3), (1, 2, 2), (1, 3, 1), (2, 1, 2), (2, 2, 1), and (3, 1, 1).

17. No number is used more than once and order is important so we use permutations. The number of possible lock combinations is $P_{16,3}=3360$.

CHAPTER 3 TEST

1. If the intersection of the two events is the empty set, it means that the two events cannot both occur. Therefore, let A be the event that you were born in 1975 and let B be the event that you were born in 1985.

2. If the intersection of sets A and B is the same as same A, then set A must be contained in the set B. Let A be the set of all Toyota Camrys and let B be the set of all Toyotas.

3. (a) $C_{17,4}=2380$ (b) $P_{17,4}=57{,}120$

4. A countably infinite set is one that can be put into a one-to-one relationship with the natural numbers. For the set {7, 11, 15, 19, 23, …}, we have the following.

$$
\begin{array}{cccccc}
1 & 2 & 3 & 4 & 5 & \ldots & k \\
\updownarrow & \updownarrow & \updownarrow & \updownarrow & \updownarrow & & \updownarrow \\
7 & 11 & 15 & 19 & 23 & \ldots & 3+4^k
\end{array}
$$

5. $x + 85 = 190$ $\qquad$ $y + x + 105 = 360$

    $x = 105$ $\qquad$ $y + 85 + 105 = 360$

    $\qquad\qquad\qquad$ $y = 170$

    Thus, 170 students are taking only an English class.

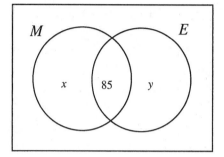

6. (a) $5 \subset I$ is not correct since 5 is not a set. Use $5 \in I$.
    (b) $5 \in F$ is correct.
    (c) $F \notin I$ is not correct since F is not an element of I. Use $F \subset I$.
    (d) $4 \notin F$ is correct and true.

7. By the basic counting law, there are $6 \times 3 \times 4 = 72$ possible executive teams.

8. In each of the three parts, we must account for six cards.
    (a) 1056; Since we want 3 of the four aces, 2 of the four kings, and one of the 44 other cards in the deck, the number of ways is this can be done is
    $C_{4,3} \times C_{4,2} \times C_{44,1} = 1056$.
    (b) 12,672; Since we want 3 of the four aces, 2 of four cards in any of the 12 other ranks, and one of the 44 other cards in the deck, the number of ways this can be done is $C_{4,3} \times \left(12 \times C_{4,2}\right) \times C_{44,1} = 12,672$.
    (c) 164,736; Since we want 3 of four cards in any of the 13 ranks, 2 of four cards in any of the 12 other ranks, and one of the 44 other cards in the deck, the number of ways this can be done is $\left(13 \times C_{4,3}\right) \times \left(12 \times C_{4,2}\right) \times C_{44,1} = 164,736$.

9.

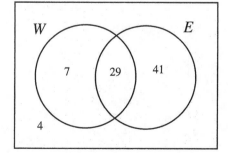

10. There are $41 + 29 + 7 = 77$ students either working more than 10 hours per week or enrolled in more than 6 units.

11. There $4 + 7 + 41 = 52$ students who are enrolled in 6 or fewer units or are working 10 or fewer hours per week.

12. There 4 students who are enrolled in 6 or fewer units and are working 10 or fewer hours per week.

CHAPTER 4 PROBABILITY

In this chapter, you will examine some basic components of probability. You will learn techniques that determine the chance that an event will occur and the corresponding odds. You will also discover that, by looking at the probabilities, the amount wagered, and the amount won or lost on the wager, you can determine if a game is a fair game. If you like to play games of chance, you will find that this chapter will make you a more knowledgeable player.

SECTION 4.1 INTUITIVE CONCEPTS OF PROBABILITY

This section discusses probability and odds. The basic formula used to determine the **probability** of an event E is given by:

$$P(E) = \frac{\text{number of ways an event can occur (\textit{desired})}}{\text{total number of possible outcomes (\textit{total})}} \quad \text{where } 0 \le P(E) \le 1.$$

The probability of the nonoccurrence of the event (its complement $\overline{E}$) is given by:

$$P(\overline{E}) = 1 - P(E).$$

For example, if E is the event of guessing the correct answer on a multiple choice question with five possible answers, the probability of guessing the correct answer is $P(E) = \frac{1}{5}$ and not guessing the correct answer is $P(\overline{E}) = 1 - \frac{1}{5} = \frac{4}{5}$.

The word probability can have two meanings – experimental probability or theoretical probability. **Experimental probability** refers to the situation in which the probability of an event has been approximated by analyzing the results of a number of trials rather than by using mathematical calculations. **Theoretical probability** refers to the situation in which the probability of an event has been predicted through mathematical calculations rather than by an experiment. The **Law of Large Numbers** states that as the number of trials increases the experimental probability approaches the theoretical probability.

Explain

1. Probability is a measure of the chance that an event will occur.

3. $P(E) = 1$ indicates that the event has a 100% chance of occurring. Tossing two standard dice and getting a sum that is less than 13 is such an event.

5. (a) Experimental probability – There is no mathematical way to determine if a serve will hit the net or not. Statistics must be kept on the player's serve to determine this.
    (b) Theoretical probability – Desired outcomes and total outcomes can be mathematically determined.
    (c) Experimental probability – There is no mathematical way to determine if a senior citizen will be on the bus. Statistics of Friday bus riders must be kept in order to determine this.
    (d) Theoretical probability – Desired outcomes and total outcomes can be mathematically determined.

7. You could use experimental methods and call a number of households to approximate the probability that a household would be watching the 10:00 P.M. news on Wednesday.

9. If E is an event, then its complement, $\overline{E}$, is the nonoccurrence of event E. The probability of the complement $\overline{E}$ is given by $P(\overline{E}) = 1 - P(E)$. For example, if E is the event of guessing the correct answer on a multiple choice question with five possible answers, the probability of guessing the correct answer is $P(E) = \frac{1}{5}$ and not guessing the correct answer is $P(\overline{E}) = 1 - \frac{1}{5} = \frac{4}{5}$.

11. If the election were held 100 times, the candidate would expect to win 63 times.

Apply

13. (a) 1/52   There is one 10 of diamonds out of the 52 cards in the deck.
    (b) 1/13   There are four tens out of the 52 cards. (4/52 = 1/13)
    (c) 10/13   There are 40 non-face cards our of the 52 cards. (40/52 = 10/13)
    (d) 1/4   There are 13 spades out of the 52 cards. (13/52 = 1/4)
    (e) 1/2   There are 26 black cards out of the 52 cards. (26/52 = 1/2)
    (f) 2/13   There are 4 queens and 4 kings or 8 desired outcomes out of 52 total outcomes. (8/52 = 2/13)

15. 195

$$\text{Let } x = \text{the number of hits expected}$$
$$\frac{367}{1000} = \frac{x}{530}$$
$$1000x = 194{,}510$$
$$x \approx 195$$

17. Let E = making a three-point shot. Therefore, $P(E) = 0.15$.
$$P(\overline{E}) = 1 - P(E) = 1 - 0.15 = 0.85$$

19. Let E = Martina passing her next test. Therefore, $P(E) = 0.60$.
$$P(\overline{E}) = 1 - P(E) = 1 - 0.60 = 0.40$$

Explore

21. The deck now contains $5 \times 52 = 260$ cards and five of each card.
    (a) 5/260 = 1/52   There are five 7's of spades out of the 260 cards.
    (b) 20/260 = 1/13   There are twenty 7's in the deck.
    (c) 60/260 = 3/13   There are $5 \times 12 = 60$ face cards in the deck.
    (d) 65/260 = 1/4   There are 65 hearts.
    (e) 40/260 = 2/13   There are 20 Aces and 20 Eights or 40 desired outcomes.
    (f) 19/259   There are a total of 20 Aces but one has been removed, leaving 19 aces available. Similarly, there are 259 cards available.
    (g) 18/258 = 9/129   There are a total of 20 Aces but two have been removed, leaving 18 aces available. Similarly, there are 258 cards available.

23. Let E = gene mutates. Therefore, $P(E) = 0.00006$.

$P(\overline{E}) = 1 - P(E) = 1 - 0.00006 = 0.99994$

25. (a) 8/25   Since there are 24 numbers on a card, the probability is 24/75 = 8/25.
    (b) 1/75   There is one desired outcome (G 59) out of 75 possible outcomes.
    (c) 4/75   There are four numbers in the N-row out of 75 possible outcomes.
    (d) 12/37 There is now 24 desired outcomes out of 74 possible outcomes. The probability is 24/74 = 12/37.
    (e) 24/73 There is now 24 desired outcomes out of 73 possible outcomes. The probability is 24/73.
    (f) 1/55   There is one desired outcome, the number in the B-row, out of 55 numbers that are left to be called.

SECTION 4.2   CALCULATING PROBABILITIES

This section uses the techniques discussed in Section 3.4 and Section 4.1 to investigate probabilities involved in real world situations. Some of the problems will involve using permutations and combinations while other still require actually listing the outcomes of an event or using the basic counting law.

Explain

1. (a) The basic counting law is used when items are being counted with replacement.
   (b) Permutations are used when items are counted without replacement and order matters.
   (c) Combinations are used when items are counted without replacement and order does not matter.
   (d) The listing method is used if the basic counting law, permutations, and combinations can not be used to determine the number of ways an event can occur.

3. In the game of Keno, there are only 20 winning numbers.

Apply

5. (a) 1/216
      There are $6 \times 6 \times 6 = 216$ total ways for three dice to land and there is only one way to get a sum of eighteen, (6, 6, 6). Thus, the probability is $\frac{1}{216}$.
   (b) 1/72
      There are three ways to get a sum of four, (1, 1, 2), (1, 2, 1), (2, 1, 1). Thus, the probability is $\frac{3}{216} = \frac{1}{72}$.
   (c) 1/36,
      There are six ways to get a sum of sixteen, (6, 6, 4), (6, 4, 6), (4, 6, 6), (5, 5, 6), (5, 6, 5), (6, 5, 6). Thus, the probability is $\frac{6}{216} = \frac{1}{36}$.

7. (a) $1/649,740 \approx 0.000001539$
      Since the order in which the cards are received does not matter, the total number of five card hands is $C_{52,5}$. Since the hand consists of 1 out of the 4 aces and 4 out of the 4 kings, the probability is $(C_{4,1} \times C_{4,4})/C_{52,5} = 1/649,740 \approx 0.000001539$.

(b) $1/54,145 \approx 0.00001847$

Since the hand consists of 1 out of the 4 aces and 4 out of the 4 cards in twelve possible ranks (K, Q, J, 10, 9, 8, 7, 6, 5, 4, 3, 2), the probability is $(C_{4,\,1} \times 12 \times C_{4,\,4})/C_{52,5} = 1/54,145 \approx 0.00001847$.

9. (a) $0.002777 \approx 1/360$

Since the order in which the cards are received does not matter, the total number of five card hands from two decks is $C_{104,5}$. Since the hand consists of 3 out of the 8 kings and 2 out of the 96 cards in the deck that are not kings, the probability is $(C_{8,\,3} \times C_{96,\,2})/C_{104,5} \approx 0.002777 \approx 1/360$.

(b) $0.03610 \approx 1/27.7$

Since the hand consists of 3 out of the 8 cards in thirteen possible ranks (A, K, Q, J, 10, 9, 8, 7, 6, 5, 4, 3, 2) and 2 out of the 96 cards in the deck that are not of that rank, the probability is $(13 \times C_{8,\,3} \times C_{96,\,2})/C_{104,5} \approx 0.03610 \approx 1/27.7$.

11. $1/56$

Since the order of finish matters, we use permutations. Selecting 2 horses out of a field of 8 horses can be done is $P_{8,\,2}$ ways. Since there is only one winning pair, the probability is $1/P_{8,\,2} = 1/56$.

13. (a) $0.2272 \approx 1/4.4$

Since the order in which the numbers are drawn does not matter, we use combinations. There are a total of $C_{80,\,5}$ ways to select five numbers. Since there are 20 winning numbers, their are 60 losing numbers. The number of ways to select five losing numbers is $C_{60,\,5}$. Thus, the probability is $C_{60,\,5} / C_{80,\,5} \approx 0.2272 \approx 1/4.4$.

(b) $0.0006449 \approx 1/1551$

The number of ways to select five winning numbers is $C_{20,\,5}$. Thus the probability is $C_{20,\,5} / C_{80,\,5} \approx 0.0006449$.

(c) $0.01209 \approx 1/82.7$

The number of ways to select four winning numbers is $C_{20,\,4}$. The number of ways to select one losing number is $C_{60,\,1}$. Thus the probability is $(C_{20,\,4} \times C_{60,\,1})/C_{80,\,5} \approx 0.01209 \approx 1/82.7$.

15. (a) $1/13,983,816 \approx 0.00000007151$

Since the order in which the numbers are drawn does not matter, we use combinations. There are a total of $C_{49,\,6}$ ways to select six numbers. Since there are 6 winning numbers, the number of ways to select six winning numbers is $C_{6,\,6}$. Thus, the probability is $C_{6,\,6} / C_{49,\,6} = 1/13,983,816 \approx 0.00000007151$.

(b) $0.00001845 \approx 1/54,201$

Since there are 6 winning numbers, there are 43 losing numbers. The number of ways to select five winning numbers is $C_{6,\,5}$ and the number of ways to select one losing number is $C_{43,\,1}$. Thus the probability is $(C_{6,\,5} \times C_{43,\,1})/C_{49,\,6} \approx 0.00001845 \approx 1/54,201$.

(c) $0.0009686 \approx 1/1032$

The number of ways to select four winning numbers is $C_{6,4}$ and the number of ways to select two losing numbers is $C_{43,2}$. Thus the probability is $(C_{6,4} \times C_{43,2})/C_{49,6} \approx 0.0009686 \approx 1/1032$.

17. **8/125**

Since each wheel can stop on 10 possible positions, there are a total of $10 \times 10 \times 10 = 1000$ total possible outcomes for the three wheels of the slot machine. Since each wheel can has four 7-symbols, there are a total of $4 \times 4 \times 4 = 64$ possible ways to get a 7 on each of the three wheels. Thus, the probability is $64/1000 = 8/125$.

Explore

19. **1/190** Since the order in which the two objects are picked from the 20 objects does not matter, the total number of way this can be done is $C_{20,2}$. Further, since there is only one correct pair of objects the probability is $1/C_{20,2} = 1/190$.

21. **5/14** Since there are nine possible events from which you can select three events, this can be done in a total of $C_{9,3}$ ways. The number of ways to select two of four poetry readings and one of the five other events is $C_{4,2} \times C_{5,1}$. Thus, the probability is $\left(C_{4,2} \times C_{5,1}\right)/C_{9,3} = 5/14$.

23. **1/10** If the first digit is to be a 5, there is only one desired outcome out of a total of 10 possible digits. The probability is 1/10. The rest of the characters on the license plate have no bearing on the probability.

25. (a) **1/10**

Since the order in which the three people are selected does not matter, there are a total of $C_{30,3}$ possible trios of speakers for the first day. The number of ways you and two of the 29 other students can be picked is $C_{1,1} \times C_{29,2}$. Thus, the probability is $(C_{1,1} \times C_{29,2})/C_{30,3} = 1/10$.

(b) $(C_{1,1} \times C_{14,2})/C_{15,3} = 1/5$

On the sixth day there are 15 students left from which to select three speakers. Of these 15 students, the probability that you and two of the other 14 students will be speaking is $(C_{1,1} \times C_{14,2})/C_{15,3} = 1/5$.

(c) **1**

If you were not selected to give your speech on the first nine days, there are only three students left to be selected on the tenth day. The event is certain and the probability is 1.

27. **1/907,200**

By the basic counting law, there are a total of $9 \times 10 \times 12 \times 15 \times 8 \times 7 = 907,200$ possible "Pick Six" tickets. Since there is only one winning ticket, the probability is $1/907,200 \approx 0.000001102$.

29. **1/1,215,450 ≈ 0.0000008227**

The "free" space is in the N-column and the four other numbers in the N-column must be the first numbers called. Since there are the 75 Bingo numbers, the probability is $C_{4,4}/C_{75,4} = 1/1,215,450 \approx 0.0000008227$.

Another way of expressing probability is by the use of **odds**. If the probability of an event is $P(E)$ and the probability of the complement is $P(\overline{E})$, then odds that the event will occur given by $O(E) = \dfrac{P(E)}{P(\overline{E})}$. If the odds of an event are a to b, then the probability of the event is given by $P(E) = \dfrac{a}{a+b}$. For example, if the Braves have a 60% chance of winning, the odds that they will win are $O(E) = \dfrac{P(\text{winning})}{P(\text{not winning})} = \dfrac{0.60}{1 - 0.60} = \dfrac{0.60}{0.40} = \dfrac{3}{2} = 3:2$.

If the odds that an event occurs are a to b, then the odds against the event occurring are b to a. Thus, the odds for the Braves losing is 2 to 3.

Explain

1. Odds are another way of expressing the chance an event will occur. It can be determined by finding the ratio of the probability that an event will occur to the probability that the event will not occur.

3. House odds are given to simplify wagering on an event. For example, if the house odds for an event are 50:1, for every \$1 you wager you would win \$50 (\$51 would be given to you) if the event occurs.

5. $O(E) = \dfrac{P(E)}{P(\overline{E})}$. If the odds of an event are a to b, then the probability of the event is given by $P(E) = \dfrac{a}{a+b}$.

Apply

7. The probability is $3/5 = 0.60$.
    Since the odds are 3 to 2, there are 3 ways to win and 2 ways to lose, for a total of 5 ways. Thus, the probability of winning is $p = \dfrac{3}{3+2} = \dfrac{3}{5} = 0.60$.

9. 1:3
    Since $P(E) = 0.25$, $P(\overline{E}) = 1 - P(E) = 1 - 0.25 = 0.75$. Therefore, the odds of the event are given by $O(E) = \dfrac{P(E)}{P(\overline{E})} = \dfrac{0.25}{0.75} = \dfrac{25}{75} = \dfrac{1}{3}$ or 1:3.

11. Since the house odds are 10 to 1, there are 10 ways for the house to win and 1 way for the house to lose. Thus, the probability the house wins is $P(\text{house wins}) = \dfrac{10}{10+1} = \dfrac{10}{11}$.

    The probability that you win is the same as the probability that the houses loses. Therefore, $P(\text{you win}) = 1 - \dfrac{10}{11} = \dfrac{1}{11}$.

Explore

13. (a) $0.85 = 17/20$
    Let event E = making a 3-point shot
    $$P(\bar{E}) = 1 - P(E)$$
    $$= 1 - 0.15$$
    $$= 0.85 = \frac{85}{100} = \frac{17}{20}$$

    (b) $3 : 17$
    $$O(E) = \frac{P(E)}{P(\bar{E})} = \frac{\frac{3}{20}}{\frac{17}{20}} = \frac{3}{17} = 3 : 17$$

    (c) $17 : 3$
    $$O(\bar{E}) = \frac{P(\bar{E})}{P(E)} = \frac{\frac{17}{20}}{\frac{3}{20}} = \frac{17}{3} = 17 : 3$$

15. (a) $1/28{,}561$
    House odds of $28{,}560 : 1$, indicate that the odds of winning
    are $1 : 28{,}560$. Thus, $P = \dfrac{1}{28{,}560 + 1} = \dfrac{1}{28{,}561}$.

    (b) $1 : 28{,}560$

    (c) $\$57{,}120$
    Let x = amount you should win
    $$\frac{28{,}560}{1} = \frac{x}{2}$$
    $$x = 57{,}120$$

17. (a) $2/5 = 0.4$

    House odds of $3 : 2$ are the odds for Williams losing.
    Thus, her odds of winning are $2 : 3$.

    If the odds are $2 : 3$, then $P(\text{wins}) = \dfrac{2}{2+5} = \dfrac{2}{5} = 0.4$.

    (b) $3/5 = 0.6$

    The odds of her losing are $3 : 2$, then $P(\text{losing}) = \dfrac{3}{3+2} = \dfrac{3}{5} = 0.6$.

    (c) $\$15$
    House odds of $3 : 2$, indicate that you win \$3 for each \$2 wagered.
    Let x = amount you win on a \$100 bet
    $$\frac{3}{2} = \frac{x}{10}$$
    $$2x = 30$$
    $$x = 15$$

19. (a) The probability of the gene not mutating is $1 - 0.00006 = 0.99994$.

(b) The odds of the gene mutating are $O(\text{mutating}) = \dfrac{0.00006}{0.99994} = \dfrac{6}{99,994} = \dfrac{3}{49,997}$ or 3:49,997.

(c) The odds of the gene not mutating are 49,997:3.

SECTION 4.4 PROBABILITY OF COMPOUND EVENTS

Compound events are events that consist of one or more single events. The presence of the words "or" or "and" is a strong indication of a compound event. There are several methods for calculating the probability of compound events, including the use of tables, the **Union Formula:** $P(A \cup B) = P(A) + P(B) - P(A \cap B)$, Venn diagrams, and tree diagrams.

Explain

1. A compound event is an event that consists of more than one single event. An example of a compound event is "*you drove to school and it was raining.*"

3. The probability of $P(A \text{ or } B)$ is the chance of at least one of the events A and B occurring.

5. When finding the intersection, both events must occur.

Apply

7. Using the union formula, we have
$$P(A \cup B) = P(A) + P(B) - P(A \cap B) = 0.56 + 0.48 - 0.12 = 0.92.$$

9. Using the union formula, we have the following.
$$P(A \cup B) = P(A) + P(B) - P(A \cap B)$$
$$0.86 = 0.56 + 0.48 - P(A \cap B)$$
$$P(A \cap B) = 0.56 + 0.48 - 0.86$$
$$P(A \cap B) = 0.18$$

11. The chance of B occurring is 0.24. Therefore the chance of $A \cap B$ occurring must be less than 0.24.

13. Using the union formula, we have the following.
$$P(A \cup B) = P(A) + P(B) - P(A \cap B)$$
$$0.29 = 0.48 + 0.36 - P(A \cap B)$$
$$P(A \cap B) = 0.48 + 0.36 - 0.29$$
$$P(A \cap B) = 0.55$$

However, this is not less than $P(A)$.

15. Since there are a total of 20 slips of paper and 8 of the planes are marked "Mexico", then $P(\text{fly to Mexico}) = \dfrac{8}{20} = \dfrac{2}{5}$.

17 Since there are a total of 20 slips of paper and 5 of them are in the shape of a boat, then $P(\text{cruise}) = \dfrac{5}{20} = \dfrac{1}{4}$.

19. $P(\text{fly}) = \dfrac{15}{20}$, $P(\text{Hawaii}) = \dfrac{9}{20}$, and $P(\text{fly to Hawaii}) = \dfrac{7}{20}$. Therefore, using the union formula, we have the following.

$$P(\text{fly} \cup \text{Hawaii}) = P(\text{fly}) + P(\text{Hawaii}) - P(\text{fly} \cap \text{Hawaii})$$
$$= \dfrac{15}{20} + \dfrac{9}{20} - \dfrac{7}{20} = \dfrac{17}{20}$$

21. $P(\text{cruise}) = \dfrac{5}{20}$, $P(\text{Mexico}) = \dfrac{11}{20}$, and $P(\text{cruise to Mexico}) = \dfrac{3}{20}$. Therefore, using the union formula, we have the following.

$$P(\text{cruise} \cup \text{Mexico}) = P(\text{cruise}) + P(\text{Mexico}) - P(\text{cruise} \cap \text{Mexico})$$
$$= \dfrac{5}{20} + \dfrac{11}{20} - \dfrac{3}{20} = \dfrac{13}{20}$$

23. The chance of going on a cruise to Mexico is $\dfrac{3}{20}$. Therefore, the chance that you do not go on a cruise to Mexico is $1 - \dfrac{3}{20} = \dfrac{17}{20}$.

25. The probability of having a migraine two days in a row is given by $0.25 \times 0.40 = 0.10$.

27. The chance that the Red Sox win in three games is given by the bottom branch of the tree so $p = 0.4 \times 0.4 \times 0.4 = 0.064$

29. The Red Sox can win the series in 5 games if we use any of the following orders: YYRRR, YRYRR, YRRYR, RYYRR, RYRYR, RRYYR. Each of these has a probability of $0.4^3 \times 0.6^2 = 0.02304$. Since there are 6 ways that this can be done, $p = 0.02304 \times 6 = 0.13824$.

31. The chance of the Yankees winning in 3 games was given in Example 7 as 0.216. Therefore, the chance that the Yankees do not win in 3 games is $1 - 0.216 = 0.784$.

33. The probability that the series ends in three games equals
$P(YYY) + P(RRR) = 0.6^3 + 0.4^3 = 0.28$.

35. The probability that the series takes more than 4 games is the same as the probability the series takes five games.

The chance that the Yankees win in five games can be read from the tree diagram as $6 \times 0.03456 = 0.20736$.

The chance that the Red Sox win in 5 games was given in problem 29 as 0.13824.

Thus, the probability the series ends in five games is $0.20736 + 0.13824 = 0.3456$.

Explore

37. (a)

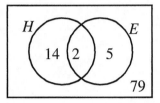

    (b)   14/100
    (c)   5/100
    (d)   79/100

39. (a)

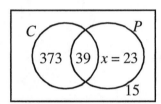

    (b)   Since there are total of 450 students in the survey and the number of students in three categories are known, we can find the number of students who use only pagers by solving the following equation $373 + 39 + x + 15 = 450$.

    This gives $x = 23$ so $P(\text{student uses only pager}) = 23/450$.
    (c) $(39 + 23)/450 = 62/450$
    (d) $(373 + 39)/450 = 412/450$

41. $(28 + 72 + 85)/730 = 185/730$

43. $150/730$

45. The probability of a student being enrolled in less than 6 units or working less than 10 hour per week is given by the union formula.

$$P(< 6 \text{ units} \cup < 10 \text{ hrs}) = P(< 6 \text{ units}) + P(< 10 \text{ hrs}) - P(< 6 \text{ units} \cap < 10 \text{ hrs})$$

$$= \frac{28 + 29 + 150}{730} + \frac{28 + 72 + 85}{730} - \frac{28}{730}$$

$$= \frac{364}{730}$$

47. The probability of two girls is $0.51 \times 0.51 = 0.2601$.

49. Having one boy and one girl can occur in the order BG or GB. Therefore, the probability of having one boy and one girl is $(0.49 \times 0.51) + (0.51 \times 0.49) = 0.4998$.

51. The probability of at least one girl is the same as the probability of not having two boys. The probability of having two boys is $0.49 \times 0.49 = 0.2401$. Therefore, the probability of having at least one girl is $1 - 0.2401 = 0.7599$.

SECTION 4.5 CONDITIONAL PROBABILITY

Conditional probability is the probability of an event occurring if some other condition has already occurred. As was done in Section 4.4, conditional probability can sometimes be found by simply reading the problem. We also use tables, tree diagrams, Venn diagrams, and the conditional probability formula.

Conditional Probability Formula:

The probability that event A occurs given that event B occurs is given by the formula:

$$P(A \mid B) = \frac{P(A \cap B)}{P(B)} \quad or \quad P(A \cap B) = P(A \mid B)P(B).$$

Explain

1. Conditional probability is the probability of an event occurring if some other condition has already occurred.

3. $P(A \mid B)$ represents the probability of a person who likes bike riding will also like to eat apples.

5. $P(A \mid B)$ is the chance that a biker rider will like eating apples. $P(A \cap B)$ is the chance of a person selected at random liking both apples and bike riding.

Apply

7. $P(A \mid B) = \dfrac{P(A \cap B)}{P(B)} = \dfrac{0.12}{0.54} = \dfrac{2}{9} \approx 0.2222$

9. $P(A \cap B) = P(A \mid B)P(B) = 0.12 \times 0.56 = 0.0672$

11. Starting with the union formula, we have the following.
$$P(A \cup B) = P(A) + P(B) - P(A \cap B)$$
$$0.86 = 0.56 + 0.48 - P(A \cap B)$$
$$P(A \cap B) = 0.56 + 0.48 - 0.86$$
$$P(A \cap B) = 0.18$$

Now, apply the conditional probability formula.
$$P(B \mid A) = \frac{P(A \cap B)}{P(A)} = \frac{0.18}{0.56} \approx 0.3214$$

13. If you know you won a plane flight, we only look at the "plane" row of the chart. Therefore, $P(\text{Mexico} \mid \text{plane}) = \dfrac{8}{15}$.

15. If you won a trip to Hawaii, we only look at the "Hawaii" column of the chart. Therefore, $P(\text{cruise} \mid \text{Hawaii}) = \dfrac{2}{9}$.

17. There are 1300 females in the survey of which 234 are in Professional Specialty. Therefore, $p = 234/1300$.

19. There are 397 people employed in a Professional Specialty and 163 are males. Therefore, $p = 163/397$.

21. (a)

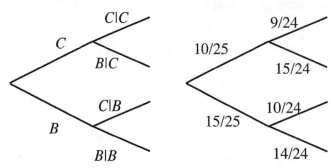

   (b) $P(C \text{ first and then B}) = P(B \mid C) \times P(C) = \dfrac{15}{24} \times \dfrac{10}{25} = \dfrac{1}{4} = 0.25$

Explore

23. (a)

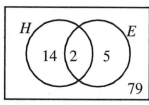

   (b) Since there were a total of 7 dogs with epilepsy and two had a serious hip problem, $p = 2/7$.

   (c) Since there were a total of 16 dogs with a serious hip problem and two had epilepsy, $p = 2/16$.

25. (a)

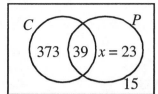

Since there are total of 450 students in the survey, and the number of students in three categories are known, we can find the number of students who use only pagers by solving the following equation $373 + 39 + x + 15 = 450$.

This gives $x = 23$ so $P(\text{student uses only pager}) = 23/450$.

(b) Since there are a total of $373 + 39 = 412$ students who use cell phones and 39 of these use pager, $p = 39/412$.

(c) There are $373 + 15 = 388$ students who do not use pagers. Of these, 373 use a cell phone. Therefore, $p = 373/388$.

27. There are 185 students working less than 10 hours per week and 85 of them are enrolled in more than 13 units. Therefore, $p = 85/185$.

29. There are 219 students enrolled in more than 13 units and 62 of them are working more than 20 hours. Therefore, $p = 62/219$.

31. If the first child is a girl, the chance of having two girls is the same as having the second child being a girl, so $p = 0.51$.

33. If the first child is a girl, the chance of having one girl and one boy is the same as having the second child being a boy, so $p = 0.49$.

SECTION 4.6 EXPECTED VALUE

This section discusses the concept of expected value. An **expected value** is the anticipated outcome of an event. This could be the anticipated amount of money won when playing a game of chance or the anticipated cost of a project based on associated risks involved in the project. If the expected value of a game is zero (0), the game is said to be a **fair game**. **Decision theory** is the area of mathematics in which decisions are made by comparing the expected values of various options. Expected values are calculated by adding the probability of each event multiplied by the amount won (lost) if the event occurs.

Explain

1. An expected value is the anticipated outcome of an event. It is calculated by adding the probability of each possibility multiplied by the amount won (lost).

3. An expected value of –$0.15 on a dollar bet means that over a long period of time the players will lose an average of $0.15 on every dollar played on the slot machine.

5. Decision theory is the area of mathematics in which decisions are made by comparing the expected values of various options.

7. The owners of the carnival game would lose money.

Apply

9. E.V. ≈ -0.3333; lose: \$33.33

$$\text{E.V.} = \frac{1}{36}(5) + \frac{9}{36}(1) + \frac{26}{36}(-1) = -\frac{12}{36} \approx -0.3333$$

Thus, you can expect to lose about \$0.3333 each time you play the game. If you played this game 100 time, you would expect to lose approximately \$33.33.

11. \$20.00

For the game to be fair, the E.V. must equal zero.
Let x = the amount the player should win
$$0.2x + 0.8(-5) = 0$$
$$0.2x - 4 = 0$$
$$0.2x = 4$$
$$x = \frac{4}{0.2} = 20$$

13. (a) E.V. ≈ 0.225

First convert the odds to probabilities using the fact that

if the odds are $a : b$, then $P = \dfrac{a}{a+b}$.

$$1 : 49 \rightarrow P = \frac{1}{1+49} = \frac{1}{50} = 0.02$$

$$1 : 999 \rightarrow P = \frac{1}{1+9999} = \frac{1}{10,000} = 0.0001$$

$$1 : 99,999 \rightarrow P = \frac{1}{1+99,999} = \frac{1}{100,000} = 0.00001$$

E.V. $= 0.02(10) + 0.0001(50) + 0.00001(2000) = 0.225$ or $22.5¢$

(b) If the cost of a stamp is greater than $22.5¢$, the contest is not worth the price of the stamp.

(c) The contest would be considered "fair" if postage is $22.5¢$.

15. ≈ 10

$$\text{E.V.} = 0.259(5) + 0.298(5) + 0.324(4) + 0.328(4) + 0.265(4)$$
$$+ 0.262(4) + 0.227(4) + 0.246(4) + 0.194(3)$$
$$= 9.975 \approx 10$$

17. ≈ -0.1407

$$\text{E.V.} = \frac{2}{2783}(\$149.50) + \frac{2}{2783}(\$74.50) + \frac{4}{2783}(\$9.50) + \frac{20}{2783}(\$2.50)$$
$$+ \frac{250}{2783}(\$0.50) + \frac{400}{2783}(0) + \frac{2105}{2783}(-\$0.50)$$
$$\approx -\$0.1407$$

Explore

19. E.V. ≈ -0.0526

Let E be the event that the ball lands on a square of four numbers

$$P(E) = \frac{4}{38} = \frac{2}{19} \quad \text{and} \quad P(\overline{E}) = 1 - \frac{2}{19} = \frac{17}{19}$$

$$\text{E.V.} = \frac{2}{19}(\$8.00) + \frac{17}{19}(-\$1.00) = 0.8421 - 0.8947 = -\$0.0526$$

21. $1.00 Ticket: E.V. ≈ -0.2664
$2.00 Ticket: E.V. ≈ -0.5326
$3.00 Ticket: E.V. ≈ -0.7990
It does not matter which amount you wager. The casino expects over 26¢ profit per dollar wagered on any of the three tickets.

$1.00 Ticket	No. of Winners	Probability		Amount Won	
	5	$\dfrac{C_{20,\,5}}{C_{80,\,5}}$	$\times$	$819	$=$ 0.5282
	4	$\dfrac{C_{20,\,4} \times C_{60,\,1}}{C_{80,\,5}}$	$\times$	$9	$\approx$ 0.1088
	3	$\dfrac{C_{20,\,3} \times C_{60,\,2}}{C_{80,\,5}}$	$\times$	$0	$\approx$ 0.0000
	2	$\dfrac{C_{20,\,2} \times C_{60,\,3}}{C_{80,\,5}}$	$\times$	$-$1	$\approx$ $-$0.2705
	1	$\dfrac{C_{20,\,1} \times C_{60,\,4}}{C_{80,\,5}}$	$\times$	$-$1	$\approx$ $-$0.4057
	0	$\dfrac{C_{60,\,5}}{C_{80,\,5}}$	$\times$	$\underline{-\$1}$	$\approx$ $\underline{-0.2272}$

E.V. ≈ -0.2664
(-26.6¢ per dollar)

$2.00 Ticket	No. of Winners	Probability		Amount Won		
	5	$\dfrac{C_{20,5}}{C_{80,5}}$	$\times$	$1638	=	1.0564
	4	$\dfrac{C_{20,4} \times C_{60,1}}{C_{80,5}}$	$\times$	$18	$\approx$	0.2177
	3	$\dfrac{C_{20,3} \times C_{60,2}}{C_{80,5}}$	$\times$	$0	$\approx$	0.0000
	2	$\dfrac{C_{20,2} \times C_{60,3}}{C_{80,5}}$	$\times$	$-$2	$\approx$	-0.5409
	1	$\dfrac{C_{20,1} \times C_{60,4}}{C_{80,5}}$	$\times$	$-$2	$\approx$	-0.8114
	0	$\dfrac{C_{60,5}}{C_{80,5}}$	$\times$	$-$2	$\approx$	-0.4544

$$\text{E.V.} \approx -0.5326$$
$$(-26.6\text{¢ per dollar})$$

$3.00 Ticket	No. of Winners	Probability		Amount Won		
	5	$\dfrac{C_{20,5}}{C_{80,5}}$	$\times$	$2457	=	1.5846
	4	$\dfrac{C_{20,4} \times C_{60,1}}{C_{80,5}}$	$\times$	$27	$\approx$	0.3265
	3	$\dfrac{C_{20,3} \times C_{60,2}}{C_{80,5}}$	$\times$	$0	$\approx$	0.0000
	2	$\dfrac{C_{20,2} \times C_{60,3}}{C_{80,5}}$	$\times$	$-$3	$\approx$	-0.8114
	1	$\dfrac{C_{20,1} \times C_{60,4}}{C_{80,5}}$	$\times$	$-$3	$\approx$	-1.2171
	0	$\dfrac{C_{60,5}}{C_{80,5}}$	$\times$	$-$3	$\approx$	-0.6816

$$\text{E.V.} \approx -0.7990$$
$$(-26.6\text{¢ per dollar})$$

23. The second procedure is best.

    For the first procedure, E. V. = 0.93(8) + 0.07(2) = 7.58.

    For the second procedure, E. V. = 0.47(15) + 0.53(2) = 8.11.

    Therefore, the second procedure will provide a greater expected lifetime.

25. (a) E.V. ≈ −0.28   (b) $50.00

The areas, probabilities, and winnings for each band are given in the chart.

region	area	probability	winnings
center	4π	4/576	10
2nd band	60π	60/576	2
3rd band	192π	192/576	0
outer band	320π	320/576	−1

$$\text{E.V.} = \frac{4}{576}(10) + \frac{60}{576}(2) + \frac{192}{576}(0) + \frac{320}{576}(-1)$$

$$\approx -\$0.28$$

For the game to be fair, E. V. = 0. If the prize for hitting the center is x, we get

$$\frac{4}{576}(x) + \frac{60}{576}(2) + \frac{192}{576}(0) + \frac{320}{576}(-1) = 0$$

$$\frac{4x - 200}{576} = 0$$

$$4x - 200 = 0$$

$$4x = 200$$

$$x = \$50.$$

CHAPTER 4 REVIEW

Review Section 4.1

1. Since there are 7 red balls and a total of 19 balls, $P(\text{red}) = 7/19$.

2. Since there are 6 sides on a standard die, there are 36 possibilities. Five of these rolls, $(2, 6), (3, 5), (4, 4), (5, 3)$, and $(6, 2)$, have a total of 8. Therefore, $p = 5/36$.

3. This is an example of empirical probability since the probability statement is made on the basis of how often a storm occurred on the coast under similar weather conditions.

4. The probability of the Brittany not having the disease is $1 - 0.024 = 0.976$.

Review Section 4.2

5. Since a race is concerned with order, we use permutations. There are a total of $P_{8,2} = 56$ possibilities and only 1 possibility is correct. Therefore, $p = 1/56$.

6. Since we are working with cards, order does not matter and we use combinations. We want to have 2 of the four kings, 2 of the four queens, and one other card. Therefore,

$$P(2 \text{ kings and 2 queens}) = \frac{C_{4,2} \times C_{4,2} \times C_{44,1}}{C_{52,5}} = \frac{1584}{2{,}598{,}960} \approx \frac{1}{1641} \approx 0.0006.$$

7. This is like problem (6) with the exception that we can pick any of 13 choices for the first pair and any of 12 choices for the second pair. Therefore,

$$P(2 \text{ pairs}) = \frac{13 \times C_{4,2} \times 12 \times C_{4,2} \times C_{44,1}}{C_{52,5}} \approx \frac{1}{10.5} \approx 0.095.$$

8. Since there are 13 cards in a suit and there are four possible suits, the number of ways to get five cards in the same suit is $4C_{13,5}$. Since we also want to pick an additional two cards, we will use $C_{39,2}$. Putting this information together, we have

$$p = \frac{4C_{13,5} \times C_{39,2}}{C_{52,7}} \approx \frac{1}{35} \approx 0.029.$$

9. There are 7 winning numbers and we pick 5. That means we also pick 2 of the 43 losing numbers. Therefore, $p = \dfrac{C_{7,5} \times C_{43,2}}{C_{50,7}} \approx \dfrac{1}{5267} \approx 0.00019$.

10. There are 20 winning numbers and we pick 8. That means we also pick 2 of the 60 losing numbers. Therefore, $p = \dfrac{C_{20,8} \times C_{60,2}}{C_{80,10}} \approx \dfrac{1}{7384.5} \approx 0.00014$.

Review Section 4.3

11. The probability of an event is a ratio of the number of successes to the total possibilities. The odds of an event are the ratio of the number successes to the number of failures.

12. Since there are 13 spades in the deck and 39 cards that are not spades, the odds are 13:39 or 1:3.

13. Since the odds are 1 to 12, the probability is $\dfrac{1}{1+12} = \dfrac{1}{13}$.

14. There are 36 possible outcomes and 3 of them, $(4,6)$, $(5,5)$ and $(6,4)$, result in a sum of 10. Therefore, there are 33 ways that the sum is not 10. This gives the odds as 3:33 or 1:11.

15. The house odds are the odds against an event happening, or the odds of you losing. If the probability of an event is $\dfrac{1}{54,145}$, the odds are 1:54,144. This means the house odds are 54,144:1.

Review Section 4.4

16.

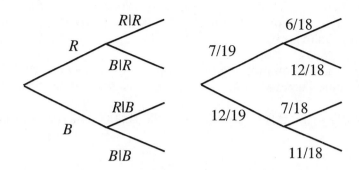

17. The chance that the first ball is red and the second one black is given by
$$p = \frac{7}{19} \times \frac{12}{18} = \frac{14}{57}.$$

18. The chance that the first ball is red or the second ball black means that all the choices, RB, RR, and BB are acceptable. The only choice that is not acceptable is BR. Since the probability of BR is given by $P(BR) = \frac{12}{19} \times \frac{7}{18} = \frac{14}{57}$,

$$P(\text{first ball red or second ball black}) = 1 - \frac{14}{57} = \frac{43}{57}.$$

19. Using the Venn diagram at the right, we have the equations
$$x + y = 120$$
$$y + z = 117 \text{ and}$$
$$x + y + z = 202 \quad (\text{from } 290 - 88).$$

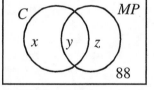

Subtracting the second equation from the third equation gives $x = 85$. Substituting $x = 85$ into the first equation gives $y = 35$. Therefore,
$$P(CD \cap MP3) = \frac{35}{290} = \frac{7}{58} \approx 0.12.$$

Review Section 4.5

20. After removing 1 black ball, the jar contains 11 black balls and 7 red balls so the probability that the next ball is black is 11/18.

21. If the first ball is returned to the jar, the jar contains 12 black balls and 7 red balls. Therefore, the probability of picking a black ball is 12/19.

22. $P(A \mid B) = \dfrac{P(A \cap B)}{P(B)} = \dfrac{0.21}{0.73} \approx 0.2877$

23. There are 204 roses, of which 82 are white. Therefore, $P(\text{White}|\text{Rose}) = 82/204$.

Review Section 4.6

24. Expected value is the average outcome of an event. This could be the expected amount of money won when playing a game or the anticipated cost of a project based on the associated risks involved in the project.

25. The expected value is found by multiplying the payoffs by the associated probabilities and then adding the results. Therefore, the expected value is given by the following.

$$EV = -1(0.8789498) + 2(0.1) + 10(0.02) + 100(0.001) + 1000(0.00005) + 100{,}000(0.0000002)$$
$$\approx -\$0.31$$

26. The game is not fair since the expected value is not equal to zero. To make the game fair, assign a value of x instead of the \$100,000 prize and set the expected value to zero.

$$0 = -1(0.8789498) + 2(0.1) + 10(0.02) + 100(0.001) + 1000(0.00005) + x(0.0000002)$$
$$0 = -\$0.3289498 + 0.0000002x$$
$$x = \frac{0.3289498}{0.0000002}$$
$$x = \$1{,}644{,}749$$

27. Computing the expected value with p being the probability your car will be completely destroyed, we have the following.

$$(4500 - 150)p + (-150)(1 - p) = 0$$
$$4350p - 150 + 150p = 0$$
$$4500p = 150$$
$$p = \frac{150}{4500} \approx 0.03$$

CHAPTER 4 TEST

1. $9/144 = 1/16$
   There are $12 \times 12 = 144$ possible rolls. Nine of these rolls, (4, 12), (5, 11), (6, 10), (7, 9), (8, 8), (9, 7), (10, 6), (11, 5) and (12, 4), have a sum of 16. Thus, the probability is 9/144.

2. The 0.5 is the theoretical probability. That does not mean that exactly 1/2 of any number of tosses will be heads. A large number of tosses will produce a probability that is approximately 0.5.

3. 1/100; Every dollar bet will give \$99 in winnings.
   The house odds of 99 : 1, means that the odds of the team winning is 1 : 99.
   Therefore, the probability of winning is $\frac{1}{1+99} = \frac{1}{100}$. The house odds of 99 : 1 indicates that each dollar bet will generate \$99 in winnings if the team is victorious.

4. $P(\text{two boys}) = 0.51 \times 0.51 = 0.2601$

5. $P(\text{two women}) = \dfrac{12}{19} \times \dfrac{11}{18} = \dfrac{132}{342} \approx 0.386$ or $\dfrac{C_{12,2}}{C_{19,2}} = \dfrac{132}{342}$

6. $P(\text{at least one women}) = 1 - P(\text{both men}) = 1 - \dfrac{7}{19} \times \dfrac{6}{18} \approx 0.877$

7. Since a race is concerned with order, we use permutations. Since there is only one way to correctly pick the top three finishers, we have $p = \dfrac{1}{P_{10,3}} = \dfrac{1}{720}$. The odds of doing this are 1:719.

8. In each of the three parts, we must account for seven cards and we will use combinations since the order of the cards does not matter. Thus, for all three parts, the denominator is given by $C_{52,7}$
    (a) Since we want 3 of the four aces and all of the four kings,
    $$P = \frac{C_{4,3} \times C_{4,4}}{C_{52,7}} = \frac{4}{C_{52,7}} = \frac{1}{33,446,140}$$
    (b) Since we want 3 of the four aces and all of some other rank,
    $$P = \frac{C_{4,3} \times 12 \times C_{4,4}}{C_{52,7}} = \frac{48}{C_{52,7}} \approx \frac{1}{2,787,178}.$$
    (c) Since we want 3 of four cards in any of the 13 ranks and all four cards in any of the 12 other ranks, $P = \dfrac{13 \times C_{4,3} \times 12 \times C_{4,4}}{C_{52,7}} = \dfrac{624}{C_{52,7}} \approx \dfrac{1}{214,398}.$

9. E.V. ≈ 0.78

roll	winnings	probability
2	–2	1/36
3	–3	2/36
4	–4	3/36
5	–5	4/36
6	–6	5/36
7	–7	6/36
8	8	5/36
9	9	4/36
10	10	3/36
11	11	2/36
12	12	1/36

    Multiplying the numbers in the winnings and probability columns and adding the results gives E. V. ≈ 0.78.

10. (a) E.V. ≈ 1.4375

Since the probability of a ball landing in each band is given by its area divided by $\sqrt{3}$, the probabilities are 1/16, 3/16 and 3/4, respectively.

This gives E.V. $= \dfrac{1}{16}(5) + \dfrac{3}{16}(2) + \dfrac{3}{4}(1) = \dfrac{23}{16} = 1.4375$.

(b) In ten tosses, the player will expect to accumulate $10 \times 1.4375 \approx 14$ points.

11. (a) $4845/1{,}581{,}580 \approx 1/326$; 1:325

Let E be the event in which you select four winning numbers.

$$P(E) = \dfrac{C_{20,\,4}}{C_{80,\,4}} = \dfrac{4845}{1{,}581{,}580} \approx \dfrac{1}{326}$$

$$O(E) = \dfrac{\frac{1}{326}}{\frac{325}{326}} = \dfrac{1}{325} = 1:325$$

(b) 325/326; 325:1

$$P(\overline{E}) = 1 - P(E) = 1 - \dfrac{1}{326} = \dfrac{325}{326}$$

$$O(\overline{E}) = \dfrac{\frac{325}{326}}{\frac{1}{326}} = \dfrac{325}{326} = 325:1$$

(c) E.V. ≈ –0.23

$$\text{E.V.} \approx \dfrac{1}{326}(\$250) + \dfrac{325}{326}(-\$1) \approx -\$.23$$

(d) Over a long period of time you can expect to lose about 23¢ on every dollar bet.

12. (a) 365^{20}

(b) $P_{365,\,20}$

(c) $P_{365,\,20} \div 365^{20} = 0.589$

(d) There are $P_{365,\,20}$ ways to select the 20 different birthdays and 365^{20} total ways to select the 20 days. Thus, the result in part (c) is the probability of selecting 20 different birthdays.

(e) If part (c) is the probability that no two of the twenty people have the same birthday, then $1 - (P_{365,\,20} \div 365^{20}) = 0.411$ is the probability that at least two of the twenty have the same birthday.

CHAPTER 5 STATISTICS AND THE CONSUMER

In this chapter, you will study some of the basic concepts of statistics. You will learn how data can be organized, displayed, and analyzed using various techniques. The focus of this chapter is on the statistics that can be applied in areas associated with consumer affairs. They will compare salaries of professional athletes, examine the margin of error in Gallup Polls and predict gasoline prices.

SECTION 5.1 ARRANGING INFORMATION

This section discusses several ways of arranging data in tabular and graphical forms so that the data is more understandable.

When data is arranged in a table, the data is first divided into categories called **classes**. The **frequencies** are the number of data values in each class. **Relative frequencies** are the frequencies divided by the total number of data values. **Percentage frequencies** are the relative frequencies, multiplied by 100. A table of data using frequencies, relative frequencies, or percentage frequencies is called a **frequency distribution**.

A frequency distribution can be represented graphically using several methods. A **bar chart** uses bars to indicate the frequency (or relative or percentage frequency) of each class. A **frequency polygon** or **line graph** uses a series of connected line segments to show the same type of information. Finally, a **pie chart** divides a circle into regions representing the percentage or relative frequencies.

When graphing data, the choice of how to graph the data is up to the person doing the analysis. In many instances, graphing the same information in different ways can emphasis different aspects of the data. For that reason it is important for the person setting up the graphs to be aware of the different forms that the graphs can take.

Explain

1.   Statistics are numerical data assembled in such a way as to present significant information about a subject.

3.   Frequencies give the number of items in each of the classes. Relative frequencies give the fraction of the total number of items that are in each class. Relative frequencies are computed by dividing the frequencies by the total number of items. Percentage frequencies express the relative frequencies as a percentage. Percentage frequencies are calculated by multiplying the relative frequencies by 100 and attaching a percent sign (%) to the results.

5.   A pie chart is a graphical representation of percentage frequencies through the uses of sectors of a circle.

7.   The angle of each sector of a pie chart is determined by multiplying the percentage for each class by 360°.

9. (a)

class	frequency
1-3	3
4-6	4
7-9	3

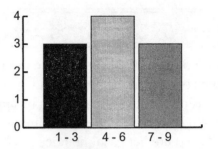

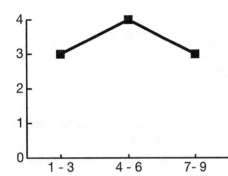

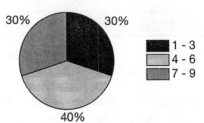

11. (a)

class	frequency
10-11	1
12-13	4
14-15	2
16-17	4
18-19	1

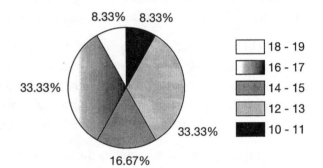

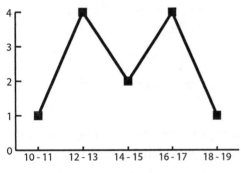

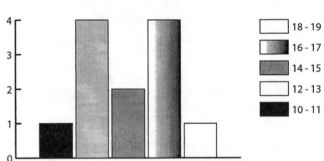

13. (a)

class	frequency
40-49	1
50-59	0
60-69	3
70-79	6
80-89	6
90-99	4

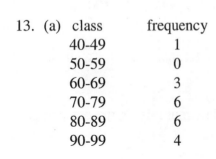

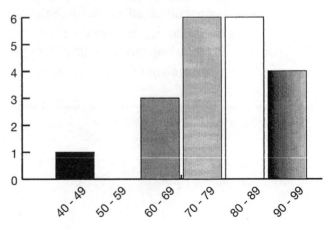

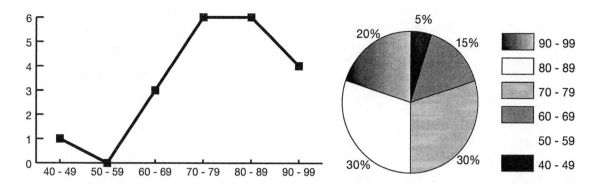

15. (a) Since $100 in 1967 was worth $512 in 2000, every $1 in 1967 was worth $5.12 in 2000. Therefore, a washing machine that cost $350 in 1967 would have cost $350 \times 5.12 = \$1792$ in the year 2000.

We could also solve this using ratios.

$$\frac{512}{100} = \frac{x}{350} \text{ so } x = 350\left(\frac{512}{100}\right) = \$1792$$

(b) Using ratios, we have the following.

$$\frac{512}{30} = \frac{40,000}{x}$$
$$512x = 30 \times 40,000$$
$$x = \frac{30 \times 40,000}{512}$$
$$x = \$2343.75$$

17. (a) Personal Expenditures in 1992 (in billions of $)

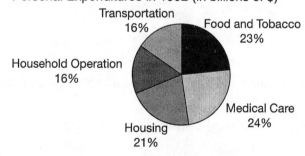

(b)

Personal Expenditures (in billions of $)

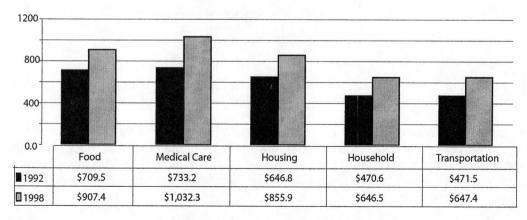

	Food	Medical Care	Housing	Household	Transportation
■ 1992	$709.5	$733.2	$646.8	$470.6	$471.5
▨ 1998	$907.4	$1,032.3	$855.9	$646.5	$647.4

(c) Medical care has increased by the largest amount ($299.1 billion) and its percentage of personal expenditures has increased from 24% to 25%.

19. (a) A pie chart would be an appropriate type of chart since it shows a comparison of the percentage of the fees that is allocated to each fund.

(b)

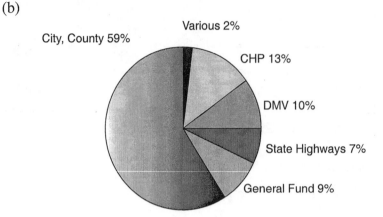

SECTION 5.2   MEASURES OF CENTRAL TENDENCY

A measure of central tendency is a number that describes the "average" of a set of data. This section discusses three such measures: the mean, the median, and the mode.

The **arithmetic mean**, μ, is the mathematical name for what many people call an average. For individual data, the formula for μ is as follows.

$$\mu = \frac{\sum_{i=1}^{n} x_i}{n}$$ where the x_i are the data values and n is the number of data values.

If the data is arranged in groups, the formula for the arithmetic mean is as follows.

$$\mu = \frac{\sum_{i=1}^{n} f_i x_i}{\sum_{i=1}^{n} f_i}$$ where $\begin{cases} x_i \text{ are the midpoints of the classes} \\ f_i \text{ are the frequencies of the classes} \\ n \text{ is the number of classes} \end{cases}$

The **median** is found by listing the data in increasing order and taking the middle value. If there are an even number of data values, the median is the mean of the two middle values.

The **mode** is the data value that occurs most frequently. If two values both have the same highest frequency, the data is said to be **bimodal**. If more than two data values have the same highest frequency, the data does not have a mode. The mode is important because it is the only measure of central tendency that can be used with non-numerical data.

Explain

1. Measures of central tendency are numbers that describe the "average" of a set of data.

3. A weighted average is the mean of a group of numbers in which certain values have more importance, or weight, than do other values. It is used when data is arranged in groups or when the importance of different values is numerically weighted.

5. If you have an odd number of data listed in increasing order, the median is the middle value. If you have an even number of data listed in increasing order, the median is the average of the two middle values.

7. The mode of a set of data is the value that occurs most frequently.

9. The mode is effective if the data is qualitative. Examples of qualitative data are the color of cars or the sizes of shoes.

Apply

11. (a) $(2 + 4 + 7 + 2 + 1 + 8 + 9 + 10 + 9 + 6)/10 = 5.8$
    (b) Listing the data in order, we have 1, 2, 2, 4, 6, 7, 8, 9, 9, 10. The median is the average of 6 and 7 = 6.5.
    (c) The data is bimodal with modes 2 and 9.

13. (a) $(20 + 29 + 2 \times 21 + 2 \times 28 + 3 \times 23 + 3 \times 26 + 5 \times 25)/17 = 24.65$
    (b) Listing the data in order, we have 20, 21, 21, 23, 23, 23, 25, 25, 25, 25, 25, 26, 26, 26, 28, 28, 29. The median is 25.
    (c) The mode is 25.

15. (a) At McDonald's

    Calories   mean $= \dfrac{280 + 330 + 470 + 340}{4} = 355$ cal

    median $= \dfrac{330 + 340}{2} = 335$ cal

    Sodium   mean $= \dfrac{590 + 830 + 890 + 890}{4} = 800$ mg

    median $= \dfrac{830 + 890}{2} = 860$ mg

    (b) At Wendy's

    Calories   mean $= \dfrac{360 + 420 + 310}{3} \approx 363$ cal

    median $= 360$ cal

    Sodium   mean $= \dfrac{580 + 920 + 790}{3} \approx 763.3$ mg

    median $= 790$ mg

17. mean = 81, median = 90, mode = 94
    The median would be the fairest method since the mode is the highest score and the mean is adversely affected by the low scores of 59 and 65.

Explore

19. (a)

	NFL	NBA	MLB	NHL
mean	$6,006,300	$14,732,000	$12,455,200	$8,254,000
median	$6,137,500	$14,900,000	$12,107,000	$8,250,000

(b) mean = $10,361,875, median = $10,680,000

(c) The median salary for the entire group is not an effective measure of the average because it does not accurate reflect the salary of the top player salaries in any particular sport.

(d) The mean salary for the entire group is not an effective measure of the average because it does not accurate reflect the salary of the top player salaries in any particular sport.

SECTION 5.3 MEASURES OF DISPERSION

A measure of dispersion describes how far data are spread from some central value. In this section, two measures of dispersion are discussed, range and standard deviation.

The **range** of a set of data is found by subtracting the lowest value from the highest value.

The standard deviation of a set of data can be found using either a standard deviation formula or a calculator.

Standard Deviation

$$\sigma = \sqrt{\frac{\sum_{i=1}^{n}(x_i - \mu)^2}{n}} \quad \text{where} \quad \begin{cases} x_i \text{ are the data values} \\ \mu \text{ is the mean} \\ n \text{ is the number data values} \end{cases}$$

There is also a formula for calculating the standard deviation of grouped data.

Standard Deviation of Grouped Data

$$\sigma = \sqrt{\frac{\Sigma f_i (x_i - \mu)^2}{n}} \quad \text{where} \quad \begin{cases} x_i = \text{the midpoint of each group} \\ f_i = \text{the frequency of each group} \\ n = \text{the total of all the frequencies} \end{cases}$$

Explain

1. A measure of dispersion describes how far data are spread from a central value.

3. The standard deviation is a measure of the dispersion of a set of data.

5. The midpoints are calculated by adding the class boundaries and dividing by 2. For example, $(20 + 29)/2 = 24.5$. The midpoints are used to estimate the values in the class

7. Since the data in experiment A has a smaller standard deviation than the data in experiment B, the results in experiment B have greater variability.

9. If a set of test scores has a standard deviation of zero, all the test scores are the same.

Apply

11. (a) 12

(b) $\mu = \dfrac{2+6+11+12+14}{5} = 9$

$$\sigma = \sqrt{\frac{(2-9)^2 + (6-9)^2 + (11-9)^2 + (12-9)^2 + (14-9)^2}{5}}$$

$$= \sqrt{\frac{49 + 9 + 4 + 9 + 25}{5}}$$

$$= \sqrt{\frac{96}{5}} \approx 4.38$$

(c) Since the new set of data is less spread out, the standard deviation is smaller.

13. (a) Range $= 8 - 0 = 8$

(b) $\mu = \dfrac{3+8+4+2+4+6+7+1+5+0}{10} = 4$

$$\sigma = \sqrt{\frac{(3-4)^2 + (8-4)^2 + (4-4)^2 + (2-4)^2 + (4-4)^2 + (6-4)^2 + (7-4)^2 + (1-4)^2 + (5-4)^2 + (0-4)^2}{10}}$$

$$= \sqrt{\frac{1 + 16 + 0 + 4 + 0 + 4 + 9 + 9 + 1 + 16}{10}}$$

$$= \sqrt{6} \approx 2.45$$

15. (a) 56   (b) 12.74

Explore

17. 132.2 mg

19. (a) Range $= 17{,}140{,}000 - 5{,}346{,}000 = \$11{,}794{,}000$

(b) \$10,680,000

(c)
salary (in millions)	# of players
\$4 - 7.999	14
\$8 - 11.999	9
\$12 - 15.999	15
\$16 - 19.999	2

(d)

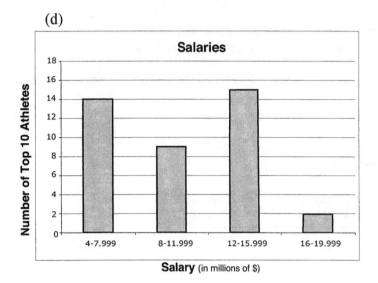

(e) $10,499,500
(f) $3,814,446

SECTION 5.4 THE NORMAL DISTRIBUTION

The normal distribution is a statistical distribution that models many real life situations. When graphed, it is a bell-shaped curve.

To convert given data into values that can be related to the standard normal distribution, we change the given data, x, into z-values using the formula

$$z = \frac{x - \mu}{\sigma} \quad \text{where} \begin{cases} x \text{ is data value} \\ \mu \text{ is the mean of the data} \\ \sigma \text{ is the standard deviation of the data} \end{cases}$$

Explain

1. The normal distribution is a statistical distribution that models many real-world situations. It is also called the bell-shaped curve.

3. The z value for a score tells the number of standard deviations between the score and the mean of the data.

5. In a normal distribution, $z = 1.53$ implies that the probability that z is between zero and 1.53 is 0.4370.

7. In a normal distribution, $z = -0.76$ implies that the area under the normal curve between $z = 0$ and $z = -0.76$ is 0.2764.

9. If $\mu = 25$, $\sigma = 5$, and $P(x > 30) = 0.159$, the probability that x is greater than 30 is 0.159.

11. 0.3413

13. $0.3413 + 0.4279 = 0.7692$

15. $0.3413 - 0.2257 = 0.1156$

17. $0.5 - 0.3665 = 0.1335$

19. $c \approx -1.675$

21. The total area is 0.6421 and the area on the left is 0.3413.
    Thus, the area on the right is $0.6421 - 0.3413 = 0.3008$. This gives $c \approx 0.845$.

23. The area to the left of $z = c$ is 0.005 so the area between $z = c$ and $z = 0$ is
    $0.5 - 0.005 = 0.4950$. Therefore, $c = -2.575$.

Explore

25. (a) $z = \dfrac{11 - 17.34}{11.14} = -0.57 \ \rightarrow \ A = 0.2157$

    $z = \dfrac{15 - 17.34}{11.14} = -0.21 \ \rightarrow \ A = 0.0832$

    Therefore, the percentage of blue collar workers with between 11 and 15 years of
    experience is $0.2157 - 0.0832 = 0.1325$ or 13.25%.

    (b) $z = \dfrac{15 - 17.34}{11.14} = -0.21 \ \rightarrow \ A = 0.0832$

    Therefore, the probability of a blue collar worker having more than 15 years of
    experience is $0.5 + 0.0832 = 0.5832$ or 58.32%.

27. (a) $z = \dfrac{20 - 17}{13} = 0.23 \ \rightarrow \ A = 0.0910$

    Therefore, the probability that a woman drives more than 20 minutes to work is
    $0.5 - 0.0910 = 0.4090$.

    (b) $z = \dfrac{20 - 17}{13} = 0.23 \ \rightarrow \ A = 0.0910$

    $z = \dfrac{10 - 17}{13} = -0.54 \ \rightarrow \ A = 0.2054$

    Therefore, the percentage of women who travel between 10 and 20 minutes to
    work is $0.0910 + 0.2054 = 0.2964$ or 29.64%.

(c) $z = \dfrac{5 - 17}{13} = -0.92 \quad \rightarrow \quad A = 0.3212$

$z = \dfrac{10 - 17}{13} = -0.54 \quad \rightarrow \quad A = 0.2054$

Therefore, the percentage of women who travel between 5 and 10 minutes traveling to work is $0.3212 - 0.2054 = 0.1158$ or 11.58%.

29. Using $A = 0.4000$, we have $z = 1.28$. This gives the following.

$$1.28 = \dfrac{x - 21.7}{20.5}$$

$$26.24 = x - 21.7$$

$$x = 47.94$$

Therefore, the 10% of men who smoke the heaviest use approximately 48 cigarettes per day.

SECTION 5.5   POLLS AND THE MARGIN OF ERROR

In this section, you will investigate confidence intervals, confidence levels, margin of error, and how these two ideas apply to public opinion polls. A **confidence interval** is an interval centered around the estimate of a statistic. A **confidence level** is a statement of the probability that the actual percentage being studied is within the confidence interval. Since a particular confidence interval will result in a particular z value, we give the common confidence levels in the following table.

Confidence Level	z value
90%	1.645
95%	1.96
98%	2.327
99%	2.575

The **margin of error** is the distance from the center of the confidence interval to each end point of the interval.

The formula for determining the margin of error is given by the following formula.

$$M = \dfrac{z}{2\sqrt{n}} \quad \text{where} \quad \begin{cases} M = \text{margin of error} \\ z = \text{value determined by the confidence} \\ \quad\quad \text{level and the normal distribution} \\ n = \text{sample size} \end{cases}$$

Explain

1.   A confidence interval is an interval centered around the estimate of a statistic.

3. The margin of error is the distance from the center of the confidence interval to each end point of the interval.

5. If the sample size increases, the margin of error decreases. As more people respond to a survey, the accuracy of the survey increases and the error decreases.

7. An estimate of a 25% viewing audience with a 5% margin of error means that, at a certain confidence level, the viewing audience is known to be between 20% and 30%.

Apply

9. $0.8000/2 = 0.4000$ so $z = 1.28$

11. $0.8500/2 = 0.4750$ so $z = 1.44$

13. $63\% - 4\% = 59\%$, $63\% + 4\% = 67\%$ so the confidence interval is 59% - 67%.

15. $47.6\% - 2.4\% = 45.2\%$, $47.6\% + 2.4\% = 50\%$ so the confidence interval is 45.2% - 50%.

17. $M = \dfrac{2.327}{2\sqrt{750}} \approx 0.042 = 4.2\%$

19. An 86% confidence level gives $z = 1.48$ so $M = \dfrac{1.48}{2\sqrt{1000}} \approx 0.023 = 2.3\%$.

21. $0.04 = \dfrac{z}{2\sqrt{2000}}$

$z = 0.04(2\sqrt{2000})$

$z = 3.58$
$z = 3.58$ so the confidence level is $2(0.49983) = 0.99966 \approx 99.97\%$.

23. $0.035 = \dfrac{z}{2\sqrt{800}}$

$z = 0.035(2\sqrt{800})$

$z = 1.98$
$z = 1.98$ so the confidence level is $2(0.4761) = 0.9522 \approx 95.22\%$.

25. $0.02 = \dfrac{2.327}{2\sqrt{n}}$

$\sqrt{n} = \dfrac{2.327}{0.04} = 58.175$

$n \approx 3385$

27. The confidence level is 80% so $z = 1.28$.

$$0.03 = \frac{1.28}{2\sqrt{n}}$$

$$\sqrt{n} = \frac{1.28}{0.06} \approx 21.3333$$

$$n \approx 456$$

Explore

29. (a) $$0.07 = \frac{z}{2\sqrt{266}}$$

$$z = 0.07(2\sqrt{266})$$

$$z = 2.28$$
$z = 2.28$ so the confidence level is $2(0.4887) = 0.9774 = 97.74\%$.

   (b) $$0.07 = \frac{2.575}{2\sqrt{n}}$$

$$\sqrt{n} = \frac{2.575}{0.14} = 18.393$$

$$n \approx 339$$

   (c) $$0.02 = \frac{2.575}{2\sqrt{n}}$$

$$\sqrt{n} = \frac{2.575}{0.04} = 64.375$$

$$n \approx 4145$$

31. (a) $$0.03 = \frac{z}{2\sqrt{1005}}$$

$$z = 0.03(2\sqrt{1005})$$

$$z = 1.90$$
$z = 1.90$ so the confidence level is $2(0.4713) = 0.9426 = 94.26\%$.

   (b) $40\% - 3\% = 37\%$, $40\% + 3\% = 43\%$ so the confidence interval is 37% to 43%.

   (c) You are 94.26% sure that the interval 37% - 43% contains the true percentage of people who thought that local economic conditions were improving.

33. (a) $$0.05 = \frac{z}{2\sqrt{506}}$$

$$z = 0.05(2\sqrt{506})$$

$$z = 2.25$$
$z = 2.25$ so the confidence level is $2(0.4878) = 0.9756 = 97.56\%$.

(b) $0.05 = \dfrac{2.575}{2\sqrt{n}}$

$\sqrt{n} = \dfrac{2.575}{0.10} \approx 25.75$

$n \approx 664$

(c) $0.04 = \dfrac{2.25}{2\sqrt{n}}$

$\sqrt{n} = \dfrac{2.25}{0.08} = 28.125$

$n \approx 792$

35. (a)

$0.03 = \dfrac{1.96}{2\sqrt{n}}$

$\sqrt{n} = \dfrac{1.96}{0.06}$

$\sqrt{n} = 32.67$

$n \approx 1068$

(b)

$0.01 = \dfrac{2.575}{2\sqrt{n}}$

$\sqrt{n} = \dfrac{2.575}{0.02}$

$\sqrt{n} = 128.75$

$n \approx 16{,}577$

SECTION 5.6 REGRESSION AND FORECASTING

In this section, you learn how to find the least-squares regression line for a set of data. While the formulas are provided in the text, it is expected that the students rely on the automated features of an advanced calculator to do the calculations. You also will look at the graph of the data and regression lines and interpret how accurately the regression models the data.

Explain

1. Linear regression is the mathematical technique for finding the best line to approximate a set of points.

3. While it always possible to compute the regression line, the line is only useful if accurately represents the data. The data should look approximately like a line.

Apply

5. $y = 1.84 + 1.05x$. When $x = 10$, $y = 1.84 + 1.05(10) = 12.34$

7. $y = 361.78 - 2.68x$. When $x = 70$, $y = 174.18$

Explore

9. (a) $y = 9.72 - 0.24x$
   (b) When $x = 34$, $y = 9.72 - 0.24(34) = 1.56$.
   (c)

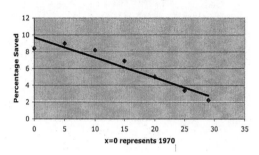

11. (a) $y = 97.04 + 14.56x$.
    (b) When $x = 25$, $y = 97.04 + 14.56(25) = \$461.04$.
    (c)

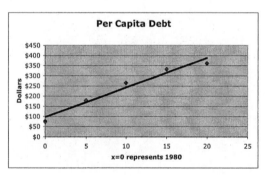

13. (a) $y = 3.78 - 0.038x$
    (b) When $x = 15$, $y = 3.78 - 0.038(15) = \$3.21$.
    (c)

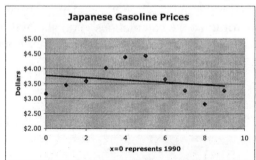

    (d) The line does not look like it gives a very accurate representation of prices.

15. $y = -641.8 + 2748.2x$

Review Section 5.1

1. Since the data values are between 1 and 50, and we want 5 classes, we use the following frequency distribution.

class	frequency
1 - 10	4
11 - 20	5
21 - 30	9
31 - 40	5
41 - 50	5

2.

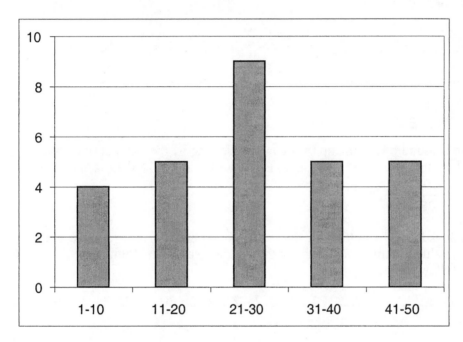

3.

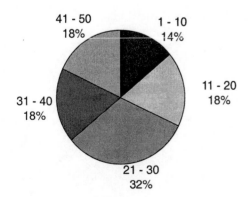

4.

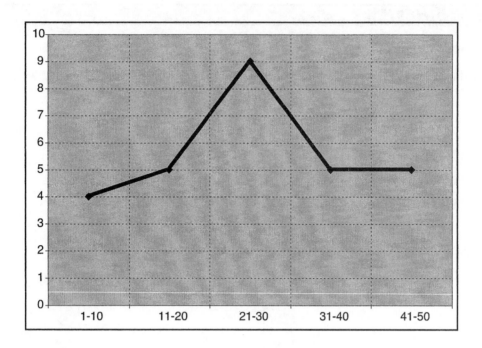

Review Section 5.2

5. The mean is found by adding up the data and dividing by the number of data values. In this problem, the sum of the data values is 708 and there are 28 data values. therefore, the mean is $\dfrac{708}{28} \approx 25.29$.

The median is found by putting the data values in order and, since we have an even number of values, taking the average of the middle two values. Putting the data in order gives the following list:

3, 4, 7, 8, 11, 12, 12, 17, 17, 21, 23, 23, 23, 25, 27, 29, 29, 30, 32, 32, 33, 36, 40, 41, 42, 43, 43, 45

Since the middle two values are 25 and 27, the median is 26.

The mode is the frequently occurring value. Since 23 occurs three times, more than any other value, the mode is 23.

6.   The mean of the frequency distribution is found by multiplying the class frequencies by the midpoints and then adding these results. The sum is then divided by the number of data values. For our problem, we have the table that follows.

class	frequency	midpoints
1 - 10	4	5.5
11 - 20	5	15.5
21 - 30	9	25.5
31 - 40	5	35.5
41 - 50	5	45.5

This gives the mean as $\mu = \dfrac{4(5.5) + 5(15.5) + 9(25.5) + 5(35.5) + 5(45.5)}{28} \approx 26.21$

7.   The mode is the best measure of central tendency when you are trying to find the most frequently occurring value. For example, if you are trying to find the most popular type of car in a parking lot, you would count the number of Fords, Toyotas, etc. and the make that appears most frequently is the mode.

8.   The median is the best measure of central tendency when you are trying to compute a numerical value but there may be a few extremely large or small values that would distort the mean. For example, if you had five house purchases in the town and the houses sold for $200,000, $225,000, $240,000, $250,000, and $2,500,000 the median would give the best measure of central tendency.

Review Section 5.3

9.   The standard deviation is best found with a statistical calculator. Doing so gives a standard deviation of approximately 12.52.

10. The standard deviation of the frequency distribution can be found by subtracting the mean form the class midpoint, squaring the result, and then multiplying that result by the frequency. adding those value, dividing by 28 and taking the square root gives the final answer.

Therefore, we have

$$\sigma = \sqrt{\dfrac{\begin{aligned}4(5.5 - 26.12)^2 + 5(15.5 - 26.12)^2 + 9(25.5 - 26.12)^2 \\ + 5(35.5 - 26.12)^2 + 5(45.5 - 26.12)^2\end{aligned}}{28}} \approx 12.80.$$

11. The range is the difference between the highest and lowest value. Therefore the range is $45 - 3 = 42$.

12. The standard deviation tells you how much the data is spread away from the mean.

Review Section 5.4

13. To find the area under the normal curve between $z = 1.5$ and $z = 2.3$, look up each value in the z table and subtract the results. If $z = 1.5$, $A = 0.4332$. If $z = 2.3$, $A = 0.4893$. Therefore the are under the normal distribution on the interval $1.5 \le z \le 2.3$ is $0.4893 - 0.4332 = 0.0561$.

14. If the area when $z \geq c$ is 0.1222, the area when $0 \leq z \leq c$ is given by $0.5000 - 0.1222 = 0.3778$. If $A = 0.3778$, $c \approx 1.16$.

15. Using $\mu = 73$ and $\sigma = 9.1$, we have $z = \dfrac{x - \mu}{\sigma} = \dfrac{90 - 73}{9.1} = 1.87$. From the normal table, we get $A = 0.4693$. Since we are looking for $P(x \geq 90)$, we have
$$P(x \geq 90) = 0.5000 - 0.4693 = 0.0307.$$

16. If the scores are normally distributed and the probability of getting an A is 12%, the probability of having a score between the mean and the lowest A is $0.50 - 0.12 = 0.38$. Looking up 0.3800 in the normal table gives $z \approx 1.175$. This gives the following.

$$z = \frac{x - \mu}{\sigma}$$

$$1.175 = \frac{x - 73}{9.1}$$

$$1.175(9.1) = x - 73$$

$$x = 73 + 1.175(9.1)$$

$$x \approx 84$$

Review Section 5.5

17. The margin of error is the distance from the center of a confidence interval to each endpoint. It is how much error is expected in the results of a survey. The confidence level is a statement that specifies the probability that the confidence interval actually contains the value being estimated.

18. The margin of error should decrease as the sample size increases since a survey with a larger number of people should give a more accurate representation of the true opinions of the group. In terms of a formula, since $M = \dfrac{z}{2\sqrt{n}}$, if n increases, the denominator of the fraction increases, forcing the value of the function to decrease.

19. If we have a 90% confidence level, $z = 1.645$. Using the formula $M = \dfrac{z}{2\sqrt{n}}$ gives the following.

$$M = \frac{z}{2\sqrt{n}}$$

$$0.01 = \frac{1.645}{2\sqrt{n}}$$

$$\sqrt{n} = \frac{1.645}{0.02} = 82.25$$

$$n \approx 6766$$

20. Using $n = 2500$ and $z = 2.327$, we get the following.

$$M = \frac{z}{2\sqrt{n}}$$

$$M = \frac{2.327}{2\sqrt{2500}}$$

$M = 0.02327$
Therefore, the margin of error is approximately 2.3%.

Review Section 5.6

21. $y = 6.69 + 1.81x$
22. If $x = 20$, $y = 6.69 + 1.81(20) = 42.89$

23. If $y = 40$, solving for x gives the following.

$$40 = 6.69 + 1.81x$$

$$1.81x = 33.31$$

$$x \approx 18.4$$

24.

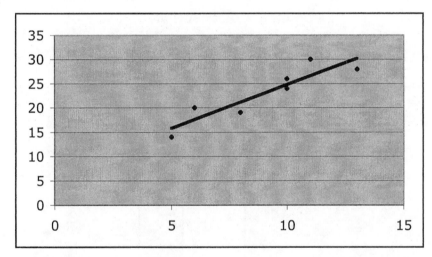

CHAPTER 5 TEST

1. (a) $\dfrac{2 + 4 + 6 + 8 + 3 + 6 + 7 + 9 + 1 + 3}{10} = 4.9$

   (b) 1, 2, 3, 3, 4, 6, 6, 7, 8, 9   median = 5
   (c) Bimodal — the modes are 3 and 6
   (d) Range = $9 - 1 = 8$
   (e) 2.55

2. (a)

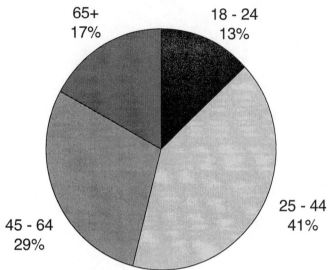

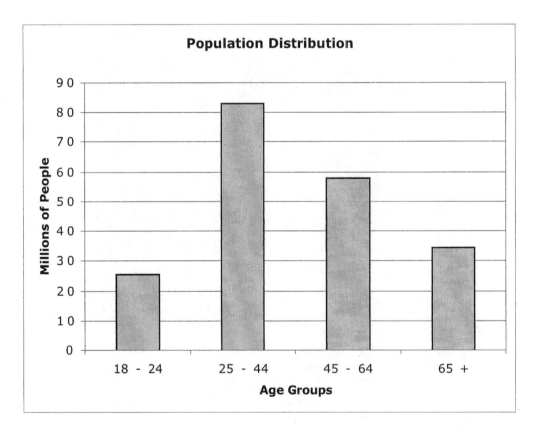

The graphs could be misleading because the age group 65+ is a much broader category than the other groups.

3. (a)

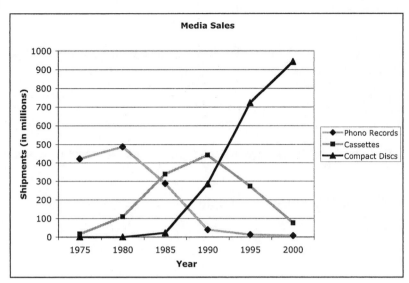

(b) Some conclusions are:
Sales of phonograph records peaked in 1980 while the sales of cassettes peaked in 1990.
Sales of compact discs increased at a rapid and fairly consistent rate since 1985.

4. (a) Mean Income (using $109,999.50 as the midpoint of the highest income category)
High School - $26,508
Some College - $33,778
Bachelors - $46,564
Advanced - $57,944

(b) Greater education provides greater income.

5. Company A because of less variation in the product.

6. (a) $z = \dfrac{20 - 18.14}{10.08} = 0.18 \quad \rightarrow \quad A = 0.0714$

$z = \dfrac{10 - 18.14}{10.08} = -0.81 \quad \rightarrow \quad A = 0.2910$

Therefore, the percentage of workers who have between 10 and 20 years of experience is $0.0714 + 0.2910 = 0.3624$ or 36.24%.

(b) $z = \dfrac{5 - 18.14}{10.08} = -1.3 \quad \rightarrow \quad A = 0.4032$

$z = \dfrac{10 - 18.14}{10.08} = -0.81 \quad \rightarrow \quad A = 0.2910$

Therefore, the percentage of men who have between 5 and 10 years of experience is $0.4032 - 0.2910 = 0.1122$ or 11.22%.

(c) $z = \dfrac{15 - 18.14}{10.08} = -0.31 \quad \rightarrow \quad A = 0.1217$

Therefore, the probability of a worker having more than 15 years of experience is $0.5 + 0.1217 = 0.6217$.

(d) $z = \dfrac{25 - 18.14}{10.08} = 0.68 \quad \rightarrow \quad A = 0.2517$

Therefore, the probability that a worker has more than 25 years of experience is $0.5 - 0.2517 = 0.2483$.

7. Using $A = 0.5000 - 0.1000 = 0.4000$, we find $z = 1.28$.
$$1.28 = \dfrac{x - 73}{14.2}$$
$$18.2 = x - 73$$
$$x = 91.2$$
Therefore, all students scoring 91.2 or higher will receive a grade of A.

Using $A = 0.5000 - 0.1000 - 0.2000 = 0.2000$, we find $z = 0.52$.
$$0.52 = \dfrac{x - 73}{14.2}$$
$$7.4 = x - 73$$
$$x = 80.4$$

Thus, students with scores between 80.4 and 91.2 will receive B's.

To find the scores for the other three grades, we will use the fact that the normal distribution is symmetric about the mean. This means we can use some of our previous work to help complete the problem.

To find the scores receiving a grade of F, we need to look at the lowest 10% of the scores. Since the diagram is the mirror image of the picture used to find the A grades, we know that the z value is $z = -1.28$.
$$-1.28 = \dfrac{x - 73}{14.2}$$
$$-18.2 = x - 73$$
$$x = 54.8$$

This means that all students scoring 54.8 or lower will receive a grade of F.

The scores earning D's, are found in a similar way. From our work on the B's, we have $z = -0.52$. This gives the following.
$$-0.52 = \dfrac{x - 73}{14.2}$$
$$-7.4 = x - 73$$
$$x = 65.6$$

This means that all students scoring between 54.8 and 65.6 will receive a grade of D.

A's 91.2-100
B's 80.4-91.19
C's 65.6-80.39
D's 54.8-65.59
F's 0-54.79

8. (a)

$$0.03 = \frac{z}{2\sqrt{1005}}$$

$z = 0.03(2\sqrt{1005}) \approx 1.90$

Since $z = 1.90$, the confidence level is $2(0.4713) = 0.94266 = 94.26\%$.

(b) $0.03 = \dfrac{2.575}{2\sqrt{n}}$

$\sqrt{n} = \dfrac{2.575}{0.06} = 42.917$

$n \approx 1842$

(c) $0.01 = \dfrac{2.575}{2\sqrt{n}}$

$\sqrt{n} = \dfrac{2.575}{0.02} = 128.75$

$n \approx 16,577$

9. The equation of the regression line is $y = 14 + 9.5x$.
    In 2010, $x = 15$ and $y = 14 + 9.5(15) = 156.5$.

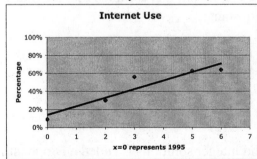

This is not a good model because the line predicts more than 100% usage after 2004.

CHAPTER 6 MODELING WITH ALGEBRA

The primary focus of math classes prior to this one has most likely been algebra. In this chapter, you will review some of the algebra from previous courses and then learn of some of the many applications of algebra. You will not investigate the typical mixture, work, interest, or motion problems of those previous classes. Instead, you will look at how algebraic functions can act as models for real life situations. You will see linear functions determine the profits at Carmen's Coffee Shop. You will find out how quadratic functions can predict the flight of a field goal in football. You will study exponential functions that estimate population growth and logarithmic functions that tell how long it takes for a vanilla latte to cool. You will see that algebra is alive in every day events and can be used to make predictions about the real world.

SECTION 6.1 LINEAR MODELS

Using the slope formula $m = \frac{y_2 - y_1}{x_2 - x_1}$ and the equation for a line $y = mx + b$, this section looks at situations in which linear functions can act as mathematical models. Linear models are appropriate when quantities change at a constant rate. If the domain of the linear function consists of real numbers, the graph is a line or a line segment. However, if the domain of the function consists only of integers, the graph consists of distinct dots lying along a straight line. Situations in which equations with integer coefficients and solutions (Diophantine Equations) will be examined at the end of the section.

Explain

1. When a linear equation is used to model a situation, it is assumed that the rate at which quantities change (the slope) is constant.

3. The value of m is the cost of producing one more item. For example, if $C = 3x + 10$, when $x = 100$, $C = 310$. If $x = 101$, $C = 313$.

5. The value of b is the initial value of the car.

7. An equation with integer coefficients and integer solutions is called a Diophantine equation.

Apply

9.

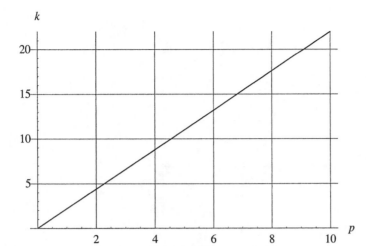

11.

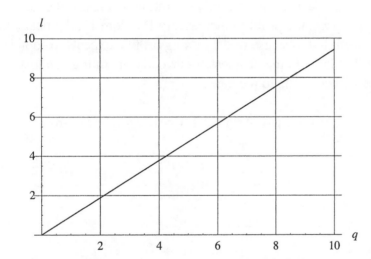

13.

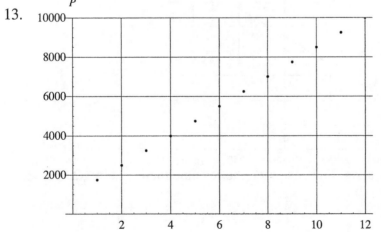

15.

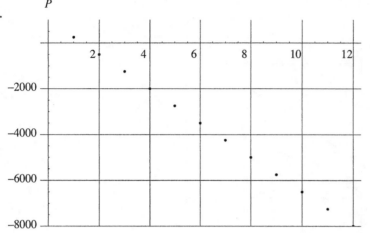

17. In Problem 13, the business is currently making a profit and expects the profit to increase. In Problem 14, the business is currently operating at a loss but expects to make a profit soon and expects the profit to increase. In Problem 15, the business is currently making a profit but, because of decreasing profits, expects to soon be operating at a loss. In Problem 16, the business is currently operating at a loss and expects that the losses will continue to increase.

19. $x = 1, y = 1$; $x = 6, y = -2$; $x = 11, y = -5$; $x = -4, y = 4$; $x = 16, y = -8$

Explore

21. (a)

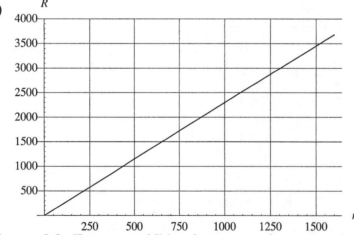

(b) $m = 2.3$; For every additional customer, the revenue increases by $2.30.

(c) $R = 2.3(1500) = \$3450$

23. (a)

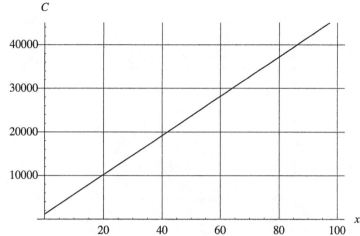

(b) Since only a whole number of barrels may be produced, the line is actually a series of points at whole number values of x.

(c) $m = 450$. This gives the cost of manufacturing one more wine barrel.

(d) $C = 450(100) + 1200 = \$46,200$

(e) $100,000 = 450x + 1200$

$98,800 = 450x$

$x = \dfrac{98,800}{450}$

$x = 219$ barrels

25. (a) Using the points of the form (a, t) and the specific points $(0, 59)$ and $(1000, 55.5)$, we have the slope given by $m = \dfrac{55.9 - 59}{1000 - 0} = -0.0035$.

Therefore the equation is given by $t = -0.0035a + 59$.

(b) $t = -0.0035(24,000) + 59 = -25°$ F

27. (a) If $B =$ number of layers of bricks and $H =$ height of the wall in inches, and using points of the form (B, H) with specific points $(0, 10)$ and $(1, 18)$, we have $m = \dfrac{18 - 10}{1 - 0} = 8$. This gives the equation of the line as $H = 8B + 10$.

(b)
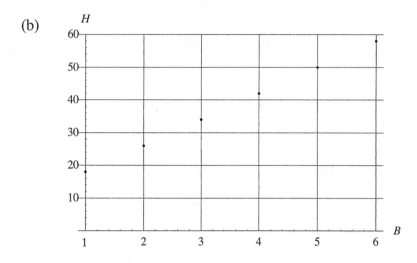

(c) To determine the height of the wall in feet, we take the height of the wall in inches and divide by 12.

$$H = \frac{8B + 10}{12}$$

$$H = \frac{8}{12}B + \frac{10}{12}$$

$$H = \frac{2}{3}B + \frac{5}{6}$$

29. (a) Using T = temperature, a = altitude, points of the form (a, T), and specific points $(0, 212)$ and $(2000, 208)$, we have $m = \dfrac{208 - 212}{2000 - 0} = -0.002$. Therefore, $T = -0.002a + 212$.

    If we use points of the form (T, a), we have the equation $a = -500T + 106,000$.

   (b) $T = -0.002(14,110) + 212 = 183.78°\text{F}$

31. (a) $T = 0.391x + 93,374$
   (b) $110,969$
$$T = 0.391x + 93,374$$
$$= 0.391(45,000) + 93,374$$
$$= 110,969$$

   (c) $365,447.19$
$$120,000 = 0.391x + 93,374$$
$$26,626 = 0.391x$$
$$68,097.19 = x$$

   Thus, taxable income is $297,350 + 68,097.19 = 365,447.19$.

33. (a) For each increase in the height, there is a constant increase in the weight.

(b) Using H = height in inches and W = weight in pounds, and points of the form (H, W), for men we have the points (67, 153) and (69, 162). This gives the following.

$$m = \frac{162 - 153}{69 - 67} = 4.5$$

$$W = mH + b$$
$$153 = 4.5(67) + b$$
$$153 = 301.5 + b$$
$$-148.5 = b$$
$$W = 4.5H - 148.5$$

Similarly, if we use the points (60, 112) and (62, 120), for women we have the equation $W = 4H - 128$.

35. 3 ways

x = number of touchdowns with extra point
y = number of field goals

$7x + 3y = 56$

The only integer solutions are $x = 2$, $y = 14$; $x = 5$, $y = 7$; $x = 8$, $y = 0$.

SECTION 6.2 QUADRATIC MODELS

Using the general quadratic function $y = ax^2 + bx + c$ and the fact that the x-coordinate of the vertex of the corresponding parabola can be found by $x = \frac{-b}{2a}$, quadratic models can be formulated to analyze various situations. In this process, solve for a, b, and c by substituting values of x and y into the general quadratic function or the vertex formula, and solving the resulting system of equations. Once a, b, and c have been determined, questions about the situation can be answered using the quadratic model.

Explain

1. The equation of a parabola is given by $y = ax^2 + bx + c$. To find the x coordinate of the vertex of a parabola, use the formula $x = \frac{-b}{2a}$. To find the y coordinate of the vertex, substitute the x value into the equation of the parabola and determine the value of y.

3. Set $y = 0$ and solve for x.

Apply

5. Vertex: $(0, 5)$, x-intercepts: $(-2.2, 0)$ and $(2.2, 0)$

    Vertex: In the equation $y = -x^2 + 5$, $a = -1$ and $b = 0$. Using the formula $x = \dfrac{-b}{2a}$ and the equation of the parabola, we have $x = \dfrac{-0}{2(-1)} = 0$ and $y = -(0)^2 + 5 = 5$. Therefore, the vertex is $(0, 5)$.

    To find the x-intercepts: If $y = 0$, then $x = \pm\sqrt{5} \approx \pm 2.2$

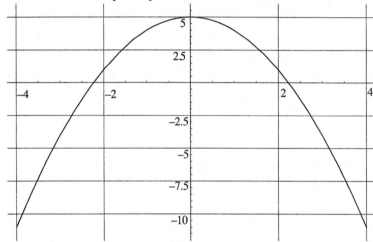

7. Vertex: $(-3, -4.7)$, x-intercepts: $(-7.9, 0)$ and $(1.0, 0)$

    Vertex: In the equation $y = 0.3x^2 + 1.8x - 2$, $a = 0.3$ and $b = 1.8$. Using the formula $x = \dfrac{-b}{2a}$ and the equation of the parabola, we have:

    $$x = \dfrac{-1.8}{2(0.3)} = -3 \text{ and } y = 0.3(-3)^2 + 1.8(-3) - 2 = -4.7.$$

    Therefore, the vertex is $(-3, -4.7)$.

    x-intercepts: If $y = 0$, using the quadratic formula gives $x = -7.0$ and 1.0.

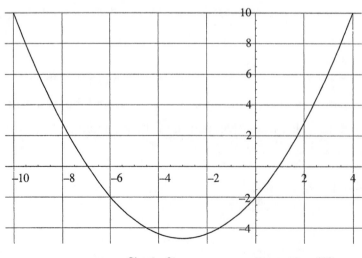

9. (a) 148.8 ft

In the equation $y = -0.042x^2 + 5x$, $a = -0.042$ and $b = 5$. Using the formula $x = \dfrac{-b}{2a}$ and the equation of the parabola, we have:

$x = \dfrac{-5}{2(-0.042)} = 59.5$ and $y = -0.042(59.5)^2 + 5(59.5) = 148.8$.

(b) 119.0 yards

Setting $y = 0$ and solving for x, we get:

$0 = -0.042x^2 + 5x$

$0 = (-0.042x + 5)x$

$x = 0$ or $-0.042x + 5 = 0$

$x = 0$ or $x = 119.0$.

11. (a) 15 ft

Set $d = 0$ and you get $h = 15$.

(b) 208.7 ft

Set $h = 0$ and solve for d.

$0 = -0.0025d^2 + 0.45d + 15$

$d = \dfrac{-0.45 \pm \sqrt{0.45^2 - 4(-0.0025)(15)}}{2(-0.0025)}$

$d = \dfrac{-0.45 \pm 0.5937}{-0.005}$

$d = 208.7$ or -28.7

(c) 35.25 ft

At the vertex of the jump,

$d = \dfrac{-(0.45)}{2(-0.0025)} = 90$

$h = -0.0025(90)^2 + 0.45(90) + 15 = 35.25$.

Explore

13. (a) $h = -0.039d^2 + 1.61d + 2$

$h = ad^2 + bd + c$
At $d = 0$, $h = 2$: $2 = a(0)^2 + b(0) + c$, therefore $c = 2$.

At $d = 35$, $h = 10$: $10 = a(35)^2 + b(35) + c \rightarrow 1225a + 42b = 8$

At $d = 42$, $h = 0$: $0 = a(42)^2 + b(42) + c \rightarrow 1764a + 42b = -2$

Solving the system of equations, we get: $a = -0.039$ and $b = 1.61$.

$$
\begin{array}{rl}
1225a + 35b = 8 & \xrightarrow{\;(6)\;} \quad 7350a + 210b = 48 \\
1764a + 42b = -2 & \xrightarrow{\;(-5)\;} \; \underline{-8820a - 210b = 10} \\
& \qquad -1470a \qquad\quad = 58
\end{array}
$$

This gives $a = -0.039$ and $b = 1.61$. Thus, $h = -0.039d^2 + 1.61d + 2$.

(b) 18.6 ft

Find the vertex of the parabola.
$$h = -0.039d^2 + 1.61d + 2$$
$$d = \frac{-b}{2a} = \frac{-1.61}{2(-0.039)} = 20.6$$
$$h = -0.039(20.6)^2 + 1.61(20.6) + 2 = 18.6$$

(c) 6.7 ft and 34.7 ft

At 11 ft above the ground, $h = 11$.
$$11 = -0.039d^2 + 1.61d + 2$$
$$0 = -0.039d^2 + 1.61d - 9$$
$$d = \frac{-1.61 \pm \sqrt{1.61^2 - 4(-0.039)(-9)}}{2(-0.039)} = \frac{-1.61 \pm 1.09}{-0.078} = 6.7 \text{ or } 34.7$$

15. (a) Let s = speed in miles per hour and D = stopping distance in feet.
We use the points $(0, 0)$, $(25, 62)$ and $(50, 195)$. Substituting the first point into the equation $D = As^2 + Bs + C$ gives
$$0 = A(0)^2 + B(0) + C \longrightarrow C = 0.$$

Substituting $C = 0$ and the other points into $D = As^2 + Bs + C$ gives

$$62 = A(25)^2 + B(25) \longrightarrow 625A + 25B = 62$$
$$195 = A(50)^2 + B(50) \longrightarrow 2500A + 50B = 195.$$

Solving this system gives $A = 0.0568$ and $B = 1.06$. This gives the equation of the parabola as $D = 0.0568s^2 + 1.06s$.

(b) At 55 mph, $D = 0.0568(55)^2 + 1.06(55) \approx 230$ feet
At 65 mph, $D = 0.0568(65)^2 + 1.06(65) \approx 309$ feet
Both of these values are slightly higher than the values given in the chart.

(c) The formula for the stopping distance is accurate to within a few feet.

(d) At 230 mph, $D = 0.0568(230)^2 + 1.61(230) \approx 3249$ feet

17. (a) $V = \dfrac{1}{2}n^2 + \dfrac{1}{2}n$

Start with the equation $V = an^2 + bn + c$ and substitute the points.
When $n = 1, V = 1$: $1 = a(1)^2 + b(1) + c \longrightarrow a + b + c = 1$

When $n = 2, V = 3$: $3 = a(2)^2 + b(2) + c \longrightarrow 4a + 2b + c = 3$

When $n = 3, V = 6$: $6 = a(3)^2 + b(3) + c \longrightarrow 9a + 3b + c = 6$

Solving the system of equations, we get: $a = \dfrac{1}{2}$, $b = 1/2$, $c = 0$.

$$\left.\begin{matrix} a + b + c = 1 \\ 4a + 2b + c = 3 \end{matrix}\right\} \quad 3a + b = 2 \xrightarrow{\;(2)\;} \quad 6a + 2b = 4$$

$$\left.\begin{matrix} a + b + c = 1 \\ 9a + 3b + c = 6 \end{matrix}\right\} \quad 8a + 2b = 5 \xrightarrow{\;(-1)\;} \quad -8a - 2b = -5$$

$$\left.\right\} -2a = -1 \longrightarrow a = \dfrac{1}{2}$$

Substituting $a = 1/2$, we get $b = 1/2$, and $c = 0$.

(b) 5050

Substituting $n = 100$ into the equation $V = \dfrac{1}{2}n^2 + \dfrac{1}{2}n$ gives:

$$V = \dfrac{1}{2}(100)^2 + \dfrac{1}{2}(100) = 5000 + 50 = 5050.$$

An exponential function contains an expression in which an exponent is a variable and its base is a constant. In this chapter, exponential functions of the form shown below will be used when finding an exponential model to fit data.

$$y = c\left(b^x\right) \text{ where } \begin{cases} b > 0 \text{ and } b \neq 1 \\ c \text{ and } k \text{ are constants} \end{cases}$$

Exponential functions are used as a model of situations in which, as the values of one variable increases at a steady rate, the values of the other variable:

1. decrease rapidly and then level off.
2. increase slowly and then increase more and more rapidly.

This section presents a variety of situations which use exponential functions as mathematical models.

Explain

1. A population tends to grow slowly when there are few people in the population and then grow more rapidly as the population increases. As long as there are no occurrences of massive war, disease, or famine, the population continues to increase. Since all parabolas decrease at some point on the curve, an exponential model is more appropriate.

3. (a) If $0 \leq b \leq 1$, as the value of x gets larger, the value of b^x gets smaller and approaches zero.
   (b) As the value of x gets smaller, the value of b^x gets larger.

5. If $b = -4$ in an exponential function $y = b^x$, we have the values $(-4)^{-2} = 1/16$, $(-4)^{-1} = -1/4$, $(-4)^0 = 1$, $(-4)^1 = -4$, and $(-4)^2 = 16$. If $x = 1/2$, we have $(-4)^{1/2} = \sqrt{-4}$ which is not a real number. Since some values of x produce imaginary numbers, the base (b) cannot be negative.

Apply

7.

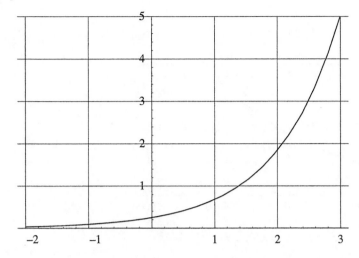

9.

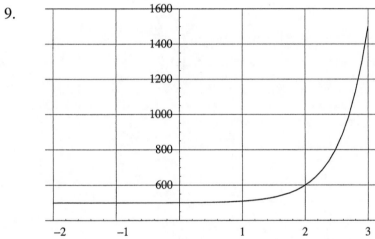

11.

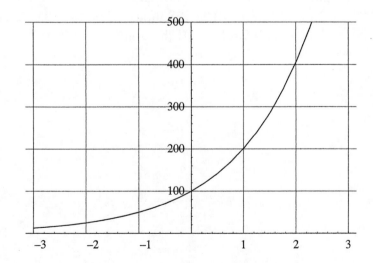

13. (a)

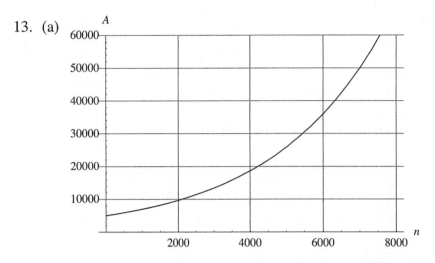

(b) $A = 5000\,(1.000328767)^{365} = \$5637.37$
(c) $A = 5000\,(1.000328767)^{20 \times 365} = \$55,094.10$

15. (a)

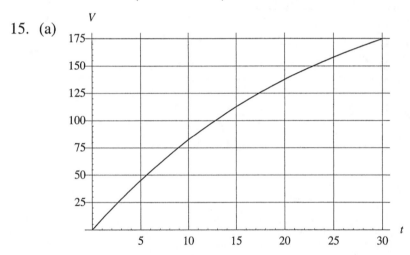

(b) $V = 250 - 250e^{-0.04\,(1)} = 250 - 240.197 = 9.803$ thousand or 9803
(c) $V = 250 - 250e^{-0.04\,(14)} = 250 - 142.802 = 107.198$ thousand or 107,198
(d) $V = 250 - 250e^{-0.04\,(30)} = 250 - 75.299 = 174.701$ thousand or 174,701

Explore

17. (a)

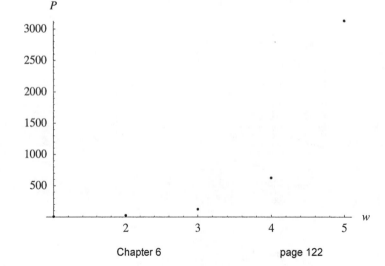

(b) Since the number of messages is always a power of 5, $p = 5^w$.

(c) $p = 5^{12} = 244{,}140{,}625$

19. 332,534,400

$$P = 282{,}000{,}000e^{(0.0201-0.0098)(16)} = 282{,}000{,}000(1.1792) = 332{,}534{,}400$$

21. 6.3 psi

$$P = 14.7(10)^{-0.000018(20{,}320)} = 6.3 \text{ psi}$$

23. $A = 407.5e^{0.1048t}$ and in 2004, $t = 12$ giving $A = 1433.2$ million.

$A = Ce^{kt}$

At $t = 0$, $A = 407.5$,

$$407.5 = Ce^{k(0)}$$
$$407.5 = C(1)$$
$$407.5 = C$$

At $t = 8$, $A = 942.5$, giving the following.

$$942.5 = 407.5e^{k(8)}$$
$$2.3129 = e^{8k}$$
$$\ln(2.3129) = \ln\left(e^{8k}\right)$$
$$\ln(2.3129) = 8k$$

$$k = \frac{\ln(2.3129)}{8} = 0.1048$$

Thus, $A = 407.5e^{0.1048t}$

In 2004, $t = 12$ and $A = 407.5e^{0.1048(12)} = 1433.2$

25. $A = 366.4e^{-0.1966t}$ and in 2004, $t = 12$ giving $A = 34.6$ million.

$A = Ce^{kt}$

At $t = 0$, $A = 366.4$ giving the following.

$$366.4 = Ce^{k(0)}$$
$$366.4 = C(1)$$
$$366.4 = C$$

At $t = 8$, $A = 76.0$ giving the following.

$$76.0 = 366.4e^{k(8)}$$
$$0.2074 = e^{8k}$$
$$\ln(0.2074) = \ln\left(e^{8k}\right)$$
$$\ln(0.2074) = 8k$$

$$k = \frac{\ln(0.2074)}{8} = -0.1966$$

Thus, $A = 366.4e^{-0.1966t}$

In 2004, $t = 12$ and $A = 366.4e^{-0.1966\,(12)} = 34.6$.

27. To determine the total number of kernels, add the values of 2^1 through 2^{64}. This gives $\approx 3.689 \times 10^{19}$ kernels of corn or $\approx 10,541,000,000,000,000$ pounds.

SECTION 6.4 LOGARITHMIC MODELS

A logarithmic function is the inverse of an exponential function. In this chapter, logarithmic functions that involve common logarithms (log) and natural logarithms (ln) will be considered. The student will graph and analyze logarithmic functions in various applications. Logarithmic functions can be used as mathematical models for situations in which a period of rapid increase or decrease is followed by a long period of gradual increase or decrease. When finding a logarithmic model to fit data, students will use logarithmic functions of the form: $y = a + b\ln(x)$ where a and b are constants.

Explain

1. The graph of $y = \log x$ has a period of rapid increase for small, positive values of x. As x becomes large, the graph has a long period of very gradual increase.

3. While initially a person may lift a small amount of weight, as the person s strength increases, the amount of weight he/she can lift increases. This will continue but the increases will become small as the person starts lifting heavier weights. This is the type of behavior exhibited by a logarithmic function.

5. Trying to take the common logarithm of a negative number with a calculator produces an error message.

Apply

7.

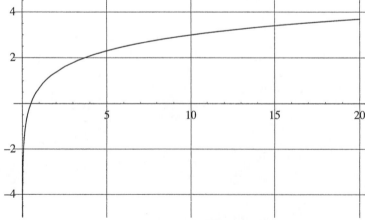

9.

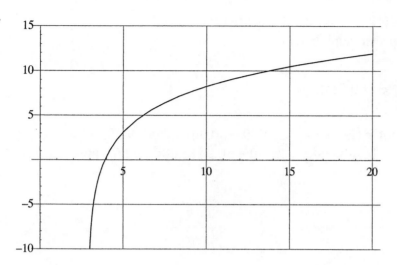

11.

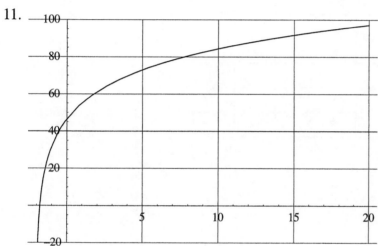

13.

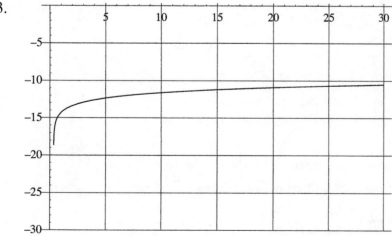

15. (a) $P = 62 + 35 \log(10 - 4) \approx 62 + 35(0.778) \approx 89.2\%$

(b) Let A = the expected adult height.
$$0.892A = 4'6'' = 54''$$

$$A = \frac{54}{0.892} = 60.5 \text{ inches}$$

17. (a) $t = 217.9 - 50.94 \ln(110 - 98) \approx 217.9 - 50.94(3.738) \approx 27.5 \text{ minutes}$

(b) $t = 217.9 - 50.94 \ln(88 - 68) \approx 217.9 - 50.94(3.000) \approx 65.3 \text{ minutes}$

(c)

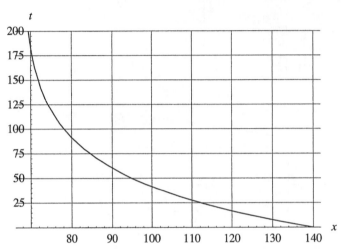

Explore

19. (a)

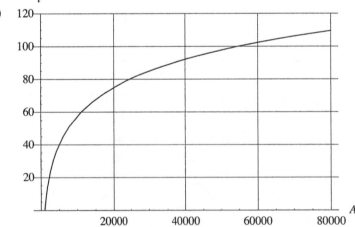

(b) $t = -172.7 + 25 \ln(1,000,000) \approx 172.7 \text{ hours}$

21. (a) $t = \dfrac{24,360}{\ln(0.5)} \ln\left(\dfrac{50}{100}\right) = 24,360 \text{ years}$

(b) $t = \dfrac{24,360}{\ln(0.5)} \ln\left(\dfrac{1}{100}\right) \approx 161,844 \text{ years}$

23. (a) The data seems to increase at a slower and slower rate.

(b) $y = 9778 + 371.7 \ln(x)$
$y = a + b\ln(x)$

At $x = 1$, $y = 9778$,

$$9778 = a + b\ln(1)$$
$$9778 = a + b(0)$$
$$9778 = a$$

At $x = 6$, $y = 10{,}444$,

$$10{,}444 = 9778 + b\ln(6)$$
$$10{,}444 = 9778 + 1.7918b$$
$$666 = 1.7918b$$
$$371.7 = b$$

Thus, $y = 9778 + 371.7\ln(x)$.

(c) When $x = 8$, $y = 10{,}551$.

In 2001, $x = 8$.
$$y = 9778 + 371.7\ln(x)$$
$$= 9778 + 371.7\ln(8)$$
$$= 10{,}551$$

The model gives an answer that is 10 less than actual amount.
($10{,}561 - 10{,}551 = 10$)

(d) 2013
$$y = 9778 + 371.7\ln(x)$$
$$10{,}900 = 9778 + 371.7\ln(x)$$
$$1122 = 371.7\ln(x)$$
$$3.0186 = \ln(x)$$
$$e^{3.0186} = x$$
$$20 = x$$

Since $x = 1$ represented 1994 and $x = 20$ is nineteen years later, the year is $1994 + 19 = 2013$.

Section 6.1

1. (a)

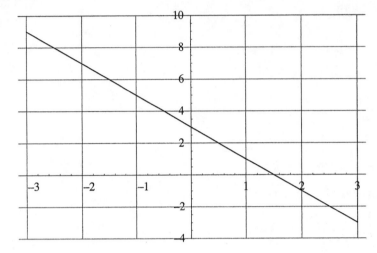

(b)

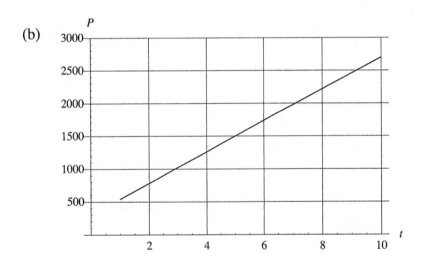

(c)

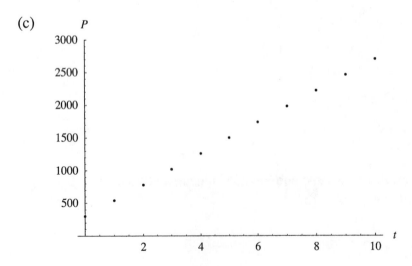

2. (a) The revenue increases at a constant rate of $0.77 for each can sold.
   (b) $P = 0.77x - 135$
   (c) $604.20

$$P = 0.77x - 135$$
$$P = 0.77(960) - 135$$
$$P = 604.20$$

   (d) 175

$$P = 0.77x - 135$$
$$0 = 0.77x - 135$$
$$135 = 0.77x$$
$$175 = x$$

   (e)
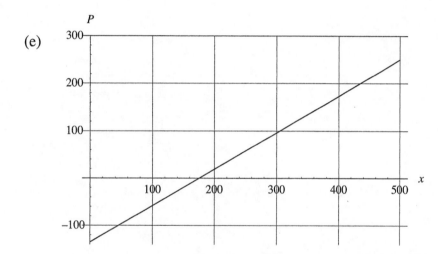

3. (a) For this problem, the points are in the form (a, s) where a = Krista s age and s = the number of stamps. Using the points $(9, 102)$ and $(14, 1567)$ we find the slope,

$$m = \frac{1567 - 102}{14 - 9} = \frac{1465}{5} = 293.$$

Substituting the slope and the point $(9, 102)$ into the slope intercept equation gives

$$s = ma + b$$
$$102 = 293(9) + b$$
$$102 = 2637 + b$$
$$b = -2535.$$

This gives the equation, $s = 293a - 2535$.

(b) $s = 293(18) - 2535 = 2739$ stamps

(c)
$$s = 293a - 2535$$
$$10,000 = 293a - 2535$$
$$12,535 = 293a$$
$$a = \frac{12,535}{293} \approx 42.8$$

So, Krista will be almost 43 years old before she has 10,000 stamps.

4. $5x + 3y = 68$, 5 ways

x = number of first-place finishers
y = number of second-place finishers
$5x + 3y = 68$

x	y
13	1
10	6
7	11
4	16
1	21

Section 6.2

5. (a) $(3, 9)$

$$y = -x^2 + 6x$$

The vertex is given by $x = \dfrac{-b}{2a} = \dfrac{-6}{2(-1)} = \dfrac{-6}{-2} = 3$ and $y = -(3)^2 + 6(3) = 9$.

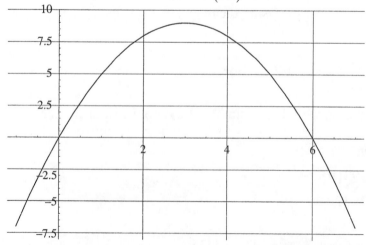

(b)   (3, –1)

$$y = x^2 - 6x + 8$$

The vertex is given by $x = \dfrac{-b}{2a} = \dfrac{-(-6)}{2(1)} = \dfrac{6}{2} = 3$ and

$$y = (3)^2 - 6(3) + 9 = -1.$$

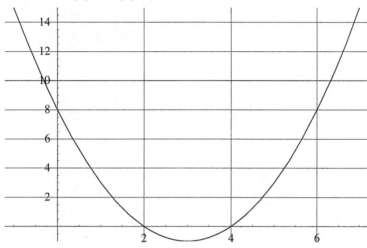

(c)   (12, 27.6)

$$y = -0.2x^2 + 4.8x - 1.2$$

The vertex is given by $x = \dfrac{-b}{2a} = \dfrac{-4.8}{2(-0.2)} = 12$ and

$$y = -0.2(12)^2 + 4.8(12) - 1.2 = 27.6.$$

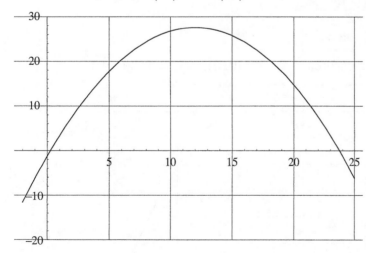

6.   (a) 7 ft
      At $x = 0$, $y = 7$.

(b) 23 ft

$$y = \frac{-x^2}{64} + x + 7$$

The vertex is given by $x = \frac{-b}{2a} = \frac{-1}{2(-1/64)} = 32$ and

$$y = \frac{-(32)^2}{64} + 32 + 7 = 23.$$

(c)

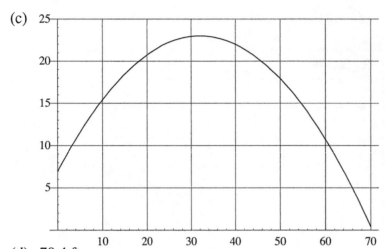

(d) 70.4 ft

$$y = \frac{-x^2}{64} + x + 7$$

$$0 = \frac{-x^2}{64} + x + 7$$

$$0 = -x^2 + 64x + 448$$

$$x = \frac{-64 \pm \sqrt{64^2 - 4(-1)(448)}}{2(-1)}$$

$$x = \frac{-64 \pm 76.7}{-2}$$

$$x = 70.4, -6.4$$

7. (a) Coordinates measured in yards: $(0, 0)$, $(98, 3)$, $(100, 0)$

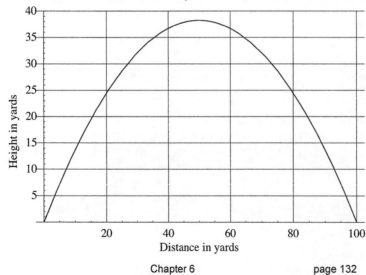

(b) $y = -0.0153x^2 + 1.53x$

$y = ax^2 + bx + c$
At $x = 0, y = 0$ → $0 = a(0)^2 + b(0) + c$ → $c = 0$
At $x = 98, y = 3$ → $3 = a(98)^2 + b(98) + c$ → $9604a + 98b = 3$
At $x = 100, y = 0$ → $0 = a(100)^2 + b(100) + c$ → $10,000a + 100b = 00$

Solving these equations simultaneously,

$$9604a + 98b = 3 \quad \xrightarrow{(100)} \quad 960,400a + 9800b = 300$$
$$10,000a + 100b = 0 \quad \xrightarrow{(-98)} \quad -980,000a - 9800b = 0$$
$$\overline{}$$
$$-19,600a = 300$$
$$a = -0.0153$$
$$b = 1.53$$

(c)   38.25 yards

$y = -0.0153x^2 + 1.53x$

The vertex is given by $x = \dfrac{-b}{2a} = \dfrac{-1.53}{2(-0.0153)} = 50$ and

$y = -0.0153(50)^2 + 1.53(50) = 38.25.$

Section 6.3

8.   (a)

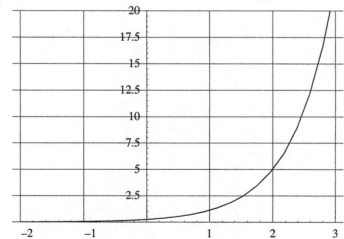

(b)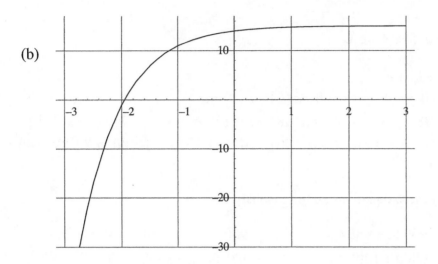

9. (a) 60 calculators

$$Q = 90 - 30e^{-0.5t}$$
$$\text{At } t = 0, \ Q = 90 - 30e^{-0.5(0)} = 60.$$

(b) 71.8, 87.5, 90.0

$$Q = 90 - 30e^{-0.5t}$$
$$\text{At } t = 1, \ Q = 90 - 30e^{-0.5(1)} = 71.8.$$
$$\text{At } t = 5, \ Q = 90 - 30e^{-0.5(5)} = 87.5.$$
$$\text{At } t = 20, \ Q = 90 - 30e^{-0.5(20)} = 90.0.$$

(c)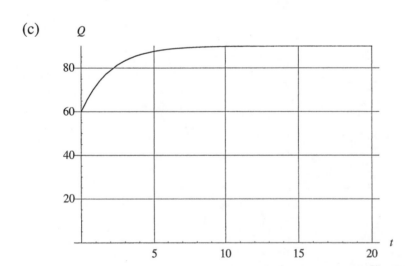

(d) 90

As the weeks of experience increases, the production approaches 90 calculators.

10. (a) The number of internet sites have increasing more and more rapidly since 1996.
    (b) $A = 0.6e^{0.9242t}$

$$A = Ce^{kt}$$

At $t = 0$, $A = 0.6$,
$$0.6 = Ce^{k(0)}$$
$$0.6 = C$$

At $t = 3$, $A = 9.6$,
$$9.6 = 0.6e^{k(3)}$$
$$16 = e^{3k}$$
$$\ln(16) = 3k\ln(e)$$
$$0.9242 = k$$

Therefore, $A = 0.6e^{0.9242t}$.

(c) The data and the results of the model agree fairly well.

t	A
0	0.6
1	1.5
2	3.8
3	9.6

(d) 24.2 million, 61.0 million
    The model predicted the number of internet sites fairly closely in 2000 but over estimated the number in 2001.

(e) The exponential model was okay for five years but it does not seem to predict the number of internet sites after the year 2000. The number of sites have stopped growing exponentially.

Section 6.4

11. (a)

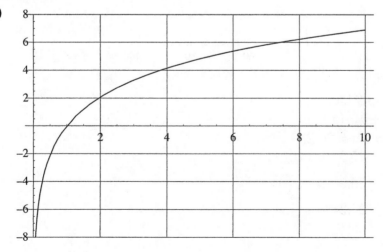

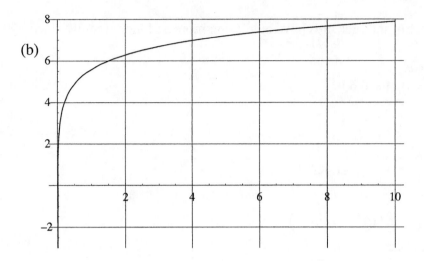

(b)

12. (a) 97, 103, 107
$p = 18 + 19.4 \ln(x)$

x	p
60	97
80	103
100	107

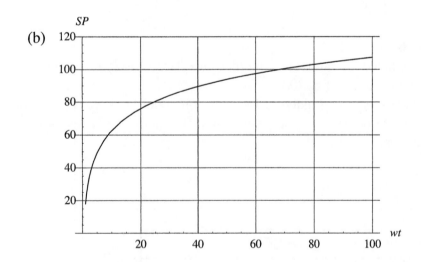

(b)

13. (a) $y = 26.2 + 13.9 \ln(x)$
$$y = a + b \ln(x)$$
At $x = 1$, $y = 26.2$,
$$26.2 = a + b \ln(1)$$
$$26.2 = a + b(0)$$
$$26.2 = a.$$

At $x = 3$, $y = 41.5$,
$$41.5 = 26.2 + b\ln(3)$$
$$15.3 = b(1.0986)$$
$$13.9 = b.$$

(b) 56.7%

In 2006, $x = 9$ and $y = 26.2 + 13.9\ln(9) = 56.7$.

(c) 2030
$$y = 26.2 + 13.9\ln(x)$$
$$75 = 26.2 + 13.9\ln(x)$$
$$48.8 = 13.9\ln(x)$$
$$3.5108 = \ln(x)$$
$$e^{3.5108} = x$$
$$33 = x$$

When $x = 33$, the year is 2030.

CHAPTER 6 TEST

1. (a)

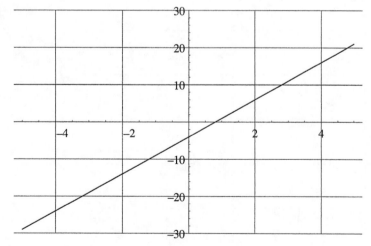

(b)

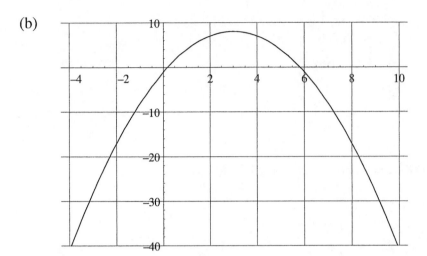

(c)

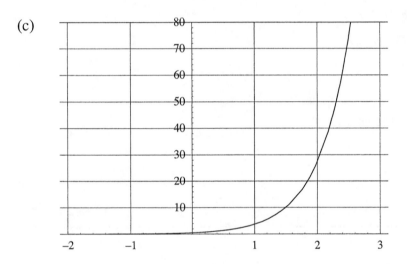

(d)

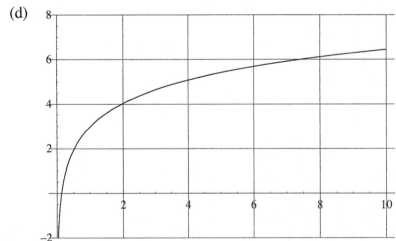

2. (a)

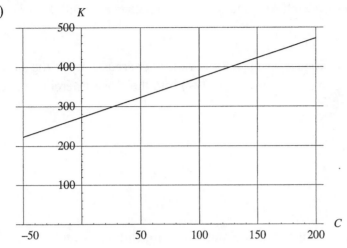

(b) $-273°C$

At absolute zero, $K = 0$. Therefore, the Celsius temperature is $-273°C$.

(c) $373°K$

If $C = 100°C$, $K = 100 + 273 = 373$.

3. (a) A constant change in the weight causes a constant change in the calories.
(b) $c = 2.727w + 140$

Using the points (110, 440) and (132, 500) we have the slope

$$m = \frac{500 - 440}{132 - 110} = \frac{60}{22} \approx 2.727.$$

Substituting the slope and the point (110, 440) into the slope intercept equation gives:
$$c = mw + b$$
$$440 = 2.727(110) + b$$
$$440 = 300 + b$$
$$b = 140$$

This gives the equation: $c = 2.727w + 140$.

(c) 576.3 cal

$$c = 2.727(160) + 140 = 576.3 \text{ cal.}$$

4. (a) The number of logs in each row changes by a constant amount.

   (b) $L = -2r + 249$

   Since the number of logs in each successive row decreases by 2, the slope is –2. Substituting the slope and the point (1, 247) into the slope intercept equation gives:

   $$L = mr + b$$
   $$247 = -2(1) + b$$
   $$247 = -2 + b$$
   $$b = 249.$$

   Thus, $L = -2r + 249$, where L is the number of logs in a row and r, the row number, is an integer with $1 \leq r \leq 124$.

   (c)

   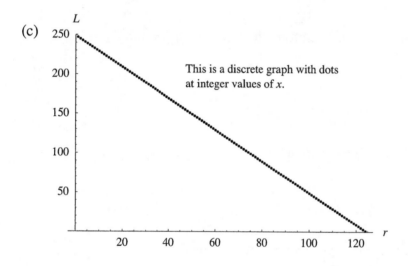

   This is a discrete graph with dots at integer values of x.

   (d) 149 logs
   At $r = 50$,
   $$L = -2r + 249$$
   $$L = -2(50) + 249$$
   $$L = -100 + 249$$
   $$L = 149.$$

   (e) 124 rows
   In the top row, $L = 1$.

   $$L = -2r + 249$$
   $$1 = -2r + 249$$
   $$-248 = -2r$$
   $$124 = r$$

5. (a)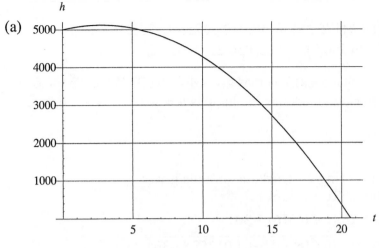

    (b) 5121 ft

    The maximum height occurs at the vertex. Using $t = \dfrac{-b}{2a} = \dfrac{-88}{2(-16)} = 2.75$, we have:

    $$h = -16(2.75)^2 + 88(2.75) + 5000 = 5121 \, \text{ft}.$$

    (c) 20.64 sec

    The rock hits the canyon floor when $h = 0$. Setting $h = 0$ and using the quadratic formula gives:

    $$0 = -16t^2 + 88t + 5000$$

    $$t = \frac{-88 \pm \sqrt{88^2 - 4(-16)(5000)}}{2(-16)}$$

    $$t \approx \frac{-88 \pm 572.49}{-32}$$

    $$t \approx 20.64 \, \text{sec}.$$

6. $3x + y = 15$; 6 ways

    Since there are three horseshoes leaning on the stake, that accounts for 6 points.
    Thus, if x is the number of 3 pointers and y is the number of 1 pointers, we get
    $3x + y = 15$ and the six solutions are (5, 0), (4, 3), (3, 6), (2, 9), (1, 12), and (0, 15).

7. (a) $h = -0.001758x^2 + 0.616x.$

    To determine the equation where x = distance of the jump and h = the height of
    the jump (both in inches), use the points (0, 0), (175.25, 54), and (350.5, 0)
    Substituting (0, 0) into the equation $h = ax^2 + bx + c$ gives:

    $$0 = a(0)^2 + b(0) + c \rightarrow c = 0.$$

    Substituting $c = 0$ and the other points into the equation $h = ax^2 + bx + c$ gives
    the following.

$$54 = a(175.25)^2 + b(175.25) \rightarrow 30,712.5625a + 175.25b = 54 \text{ and}$$
$$0 = a(350.5)^2 + b(350.5) \rightarrow 122,850.25a + 350.5b = 0$$

Solving these simultaneous equations gives $a = -0.001758$ and $b = 0.616$. Therefore, the equation modeling the jump is $h = -0.001758x^2 + 0.616x$.

(b) 4'10.5", 24'3.5"

Setting $h = 30$ and using the quadratic formula gives:
$$30 = -0.001758x^2 + 0.616x$$
$$0 = -0.001758x^2 + 0.616x - 30$$
$$x = \frac{-0.616 \pm \sqrt{0.616^2 - 4(-0.001758)(-30)}}{2(-0.001758)}$$
$$x = \frac{-0.616 \pm 0.410}{-0.00352}$$
$$x = 58.5" \text{ and } 291.5"$$
$$x = 4'\ 10.5" \text{ and } 24'\ 3.5".$$

8. (a) $P = \dfrac{3}{2}t^2 - \dfrac{1}{2}t$

Using the points (1, 1), (2, 5) and (3, 12) and substituting them into $P = at^2 + bt + c$ gives the equations:

$$1 = a(1)^2 + b(1) + c \quad \rightarrow \quad a + b + c = 1$$
$$5 = a(2)^2 + b(2) + c \rightarrow 4a + 2b + c = 5$$
$$12 = a(3)^2 + b(3) + c \rightarrow 9a + 3b + c = 12.$$

Solving these equations simultaneously gives $a = 3/2$, $b = -1/2$, and $c = 0$.

Thus, we have the equation $P = \dfrac{3}{2}t^2 - \dfrac{1}{2}t$.

(b)

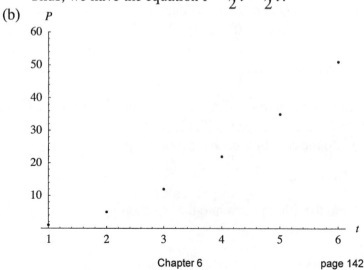

(c) 14,950

$$P = \frac{3}{2}(100)^2 - \frac{1}{2}(100) = 14,950$$

Its shape consists of a pentagon with one hundred dots on a side and 98 smaller pentagons contained inside the pentagon.

9. (a)

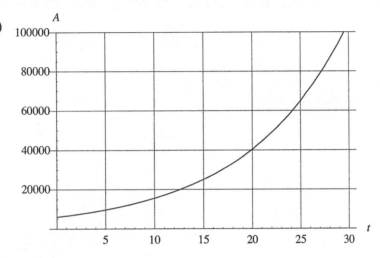

(b) $9,663.06, $15,562.45, $104,696.41
At $t = 5$, $A = 6000(1.1)^5 = 9663.06$.
At $t = 10$, $A = 6000(1.1)^{10} = 15,562.45$.
At $t = 30$, $A = 6000(1.1)^{30} = 104,696.41$.

10. (a)

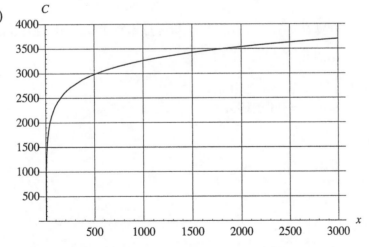

(b) $3,263.10, $3,540.36
If $x = 1000$, $y = 500 + 400 \ln(1000) = 3263.10$.
If $x = 2000$, $y = 500 + 400 \ln(2000) = 3540.36$.

(c) $3.26; \$1.77$

Taking the total costs divided by the number of trucks, we get:

$$3263.10 \div 1000 = 3.26$$
$$3540.36 \div 2000 = 1.77.$$

As the number of trucks produced increases, the cost per truck decreases.

11. (a) $y = 27.2 - 8.19 \ln(t)$ and in 2004 $(t = 7)$, $y = 11.3$.

y = number of music videos, in millions
t = time in years $(t = 1$ for 1998)

$y = a + b \ln(t)$
At $t = 1$, $y = 27.2$,

$$27.2 = a + b \ln(1)$$
$$27.2 = a + b(0)$$
$$27.2 = a.$$

At $t = 3$, $y = 18.2$,

$$18.2 = 27.2 + b \ln(3)$$
$$-9 = b(1.0986)$$
$$-8.19 = b.$$

(b) $y = 0.5e^{0.9435t}$; In 2004 $(t = 6)$, $y = 143.7$ million DVDs. (Note: the exponential model gives a very high estimates past 2003.)

y = number of DVD s, in millions
t = time in years $(t = 0$ for 1998)

$y = be^{kt}$
At $t = 0$, $y = 0.5$,

$$0.5 = be^{k(0)}$$
$$0.5 = b(1)$$
$$0.5 = b.$$

At $t = 2$, $y = 3.3$,
$$3.3 = 0.5e^{k(2)}$$
$$6.6 = e^{2k}$$
$$\ln(6.6) = \ln\left(e^{2k}\right)$$
$$1.887 = 2k\ln(e)$$
$$0.9435 = k.$$

12. (a) $A = 3^s$, where s = the square on the Monopoly board $(1 \leq s \leq 40)$ and A = amount of money.

(b) $19,683; no
If $s = 24$, $A = 3^9 = 19,683$.
$19,683 is more money than in the bank.

(c) No, more than $282,000,000,000 is needed.
If $s = 24$, $A = 3^{24} = 282,429,536,481$.

(d) $12,157,665,459,056,928,801;   303,941,636 years
If $s = 40$, $A = 3^{40} = 12,157,665,459,056,928,801$.

$12,157,665,460,000,000,000 \div \$40,000,000,000 \approx 303,941,636$ years

CHAPTER 7 GEOMETRY AND ART

In this chapter, you will learn that there is more to geometry than the definitions, postulates, theorems, proofs, and measurement formulas you may have studied in previous geometry courses. The chapter starts with a quick look at Euclidean geometry (the geometry taught from grade school through high school) and discusses the existence of non-Euclidean geometries. After that, the chapter leads you on a short journey through aspects of geometry in art such as, perspective, Golden Ratios, polygons, stars, and tessellations. The chapter ends with a look at fractal geometry, a geometry which combines the use of algebraic equations and computers to produce some interesting visual effects. You will find that you will not rehash high school geometry in this chapter. Instead, you will experience some of the artistic effects that can be produced with geometry.

SECTION 7.1 EUCLIDEAN AND NON-EUCLIDEAN GEOMETRY

This section presents a summary of some of the important definitions, axioms, postulates, and theorems of Euclidean geometry. The section includes an explanation of the need for undefined terms (point, line, plane) and postulates in the study of Euclidean geometry. The section also discusses what is meant by a theorem and the proof of a theorem. The intent of this section is to make you aware of the development of this mathematical system and to summarize some of the basic components of a system of geometry.

This section also introduces the principal non-Euclidean geometries. It shows how changing Euclid's parallel postulate can create a valid geometry with theorems that are different from those of Euclid. The two principal non-Euclidean geometries, Riemannian and Lobachevskian, are presented in this section. The chart below summarizes the basic difference between Euclidean, Riemannian, and Lobachevskian geometry.

	PARALLEL POSTULATE	**TRIANGLE SUM**	**MODEL FOR THE PLANE**
EUCLIDEAN GEOMETRY	Through a point not on a line there is **only one** parallel to a given line.	Sum of the angles of a triangle $= 180°$.	Parallelogram
RIEMANNIAN GEOMETRY	Through a point not on a line there are **no** parallels to a given line.	Sum of the angles of a triangle $> 180°$.	Sphere
LOBACHEVSKIAN GEOMETRY	Through a point not on a line there is **more than one** parallel to a given line.	Sum of the angles of a triangle $< 180°$.	Pseudosphere

The section ends with an discussion of whether the world is Euclidean, Riemannian, or Lobachevskian.

Explain

1. It is necessary to have undefined terms and postulates in a deductive system of geometry because there must be something to start the system. Not all terms can be defined nor all propositions deduced.

3. A straight angle equals 180°, a right angle equals 90°, an obtuse angle is greater than 90°, and an acute angle is less than 90°.

5. The line containing $\overline{RS}$ will intersect line m.

7. Axioms and postulates are assumptions. Axioms refer to assumptions from algebra while postulates are assumptions from geometry.

9. In Euclidean geometry, two distinct lines can intersect in at most one point.

11. In the statement of the transitive property, a and c are both equal to b. Thus, they are equal to each other ($a = c$) by Axiom A2.

13. Euclidean Triangle Sum Theorem: The sum of the angles of a triangle is 180°. Riemannian Triangle Sum Theorem: The sum of the angles of a triangle is greater than 180°. Lobachevskian Triangle Sum Theorem: The sum of the angles of a triangle is less than 180°.

Apply

15.

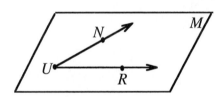

17.

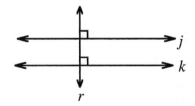

19.

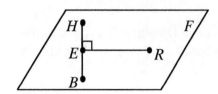

21.

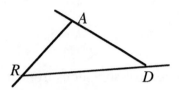

23.

Hypothesis: *m* intersects *n*
Conclusion: *m* intersects *n* in
             only one point *Q*

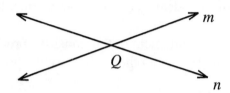

25.

Hypothesis: ∠1 is an exterior angle of ∆*ABC*
Conclusion: ∠1 > ∠*A* and ∠1 > ∠*C*

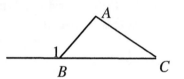

27.

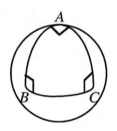

29.

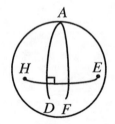

31.

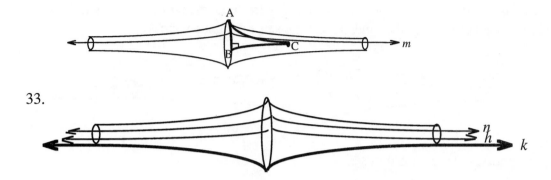

33.

Explore

35. It is not a good definition, since it does not distinguish a point from all other objects; for example, a moment in time also has no parts.

37. A possible sequence of definitions from a dictionary could be dimension $\rightarrow$ extent $\rightarrow$ length $\rightarrow$ dimension.

39.

$$5 + 6x = 4x - 11$$
$$5 + (-5) + 6x = 4x - 11 + (-5) \quad \text{(A3)}$$
$$6x = 4x - 16$$
$$6x - 4x = 4x - 4x - 16 \quad \text{(A3)}$$
$$2x = -16$$
$$x = -8 \quad \text{(A4)}$$

41. If a line intersects two parallel lines, the pairs of alternate interior angles have equal measures.

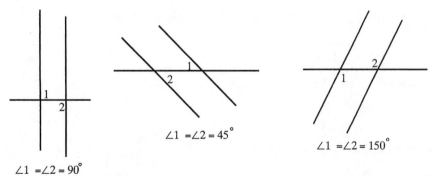

$$\angle 1 = \angle 2 = 90°$$

$$\angle 1 = \angle 2 = 45°$$

$$\angle 1 = \angle 2 = 150°$$

In all three cases, the alternating angles are equal.

43. Mathematics does not establish truths about the physical world. It can only give logical descriptions of the physical world based on beginning assumptions (postulates).

45. If you examine the set of parallel lines on the Lobachevskian model that follows, you will observe that $a \parallel b$, but as you move away from the center of the pseudosphere, the lines get closer together and are not the same distance apart.

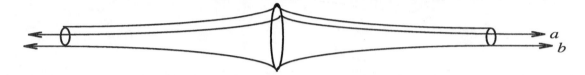

47. The fourth angle ($\angle A$) is obtuse.

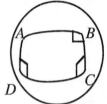

49. By drawing the diagonal of a four-sided plane figure, you get two triangles. Since in Lobachevskian geometry the sum of the angles of a triangle is less than 180°, the sum of the angles of the two triangles would be less than 360°. But a rectangle has four right angles, and the sum of its angles is exactly 360°. Thus, the Lobachevskian four-sided plane figure cannot have four right angles. Therefore, rectangles as defined do not exist in Lobachevskian geometry.

51. In equiangular $\triangle ABC$, $\angle A = \angle B = \angle C = x°$. In Lobachevskian geometry,

$$\angle A + \angle B + \angle C < 180°$$
$$x + x + x < 180°$$
$$3x < 180°$$
$$x < 60°.$$

Thus, the each angle is less than 60°.

53. In equiangular $\triangle ABC$, $\angle A = \angle B = \angle C = x°$. In Riemannian geometry,

$$\angle A + \angle B + \angle C > 180°$$
$$x + x + x > 180°$$
$$3x > 180°$$
$$x > 60°$$

Thus, the each angle is greater than 60°.

55. In Riemannian geometry, the sum of the angles is greater than 180°. Thus, it is possible for two of the three angles to be equal to 90°.

In this section, you will examine the geometry involved in drawing objects in such a way so as to give them depth and show their distance from the observer. This art of perspective drawing enables the artist to make three-dimensional objects look realistic on a two-dimensional canvas. In this section, you will examine four basic techniques of creating perspective. The method of **overlapping shapes** is a technique for creating an illusion of depth on a two dimensional surface by placing shapes in front of one another. The method of **diminishing sizes** is a technique for creating an illusion of depth on a two dimensional surface by systematically making objects smaller. **Atmospheric perspective** is a technique for creating an illusion of depth on a two dimensional surface by a gradual lessening of clarity and visual strength. **One-point perspective** is a technique for creating an illusion of depth on a two dimensional surface by the geometry of converging lines. With the geometric techniques presented in this section, you have the techniques to draw the three-dimensional world on a piece of paper.

Explain

1. Perspective is the art of drawing objects on a two-dimensional surface so as to give an illusion of depth and show distance from the observer.

3. The method of diminishing sizes is a technique for creating an illusion of depth on a two dimensional surface by systematically making objects smaller.

5. One-point perspective is a technique for creating an illusion of depth on a two dimensional surface by the geometry of converging lines.

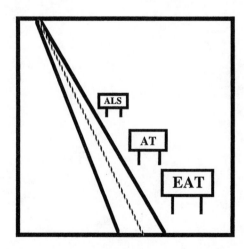

7. Draw an eye level line and a vanishing point. Objects drawn are systematically
    shortened as they recede to the vanishing point. The objects in the scene converge to
    the vanishing point by placing them within lines that recede to the vanishing point.

Apply

9. (a)

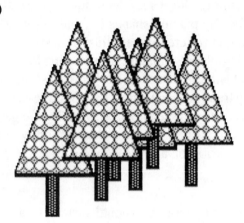

    (b)

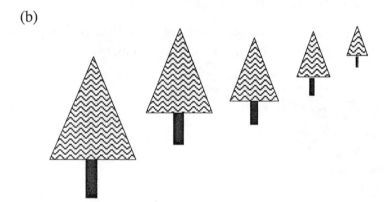

    (c)

11.

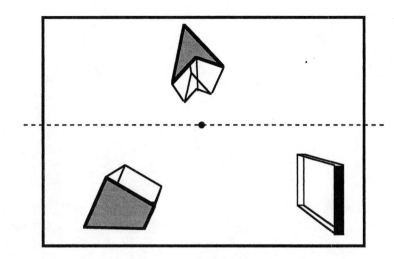

13.

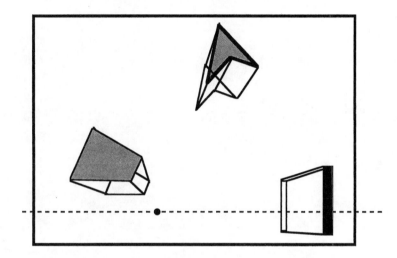

15. Use the four-step process described in Section 7.2 for equally spaced objects.

17.

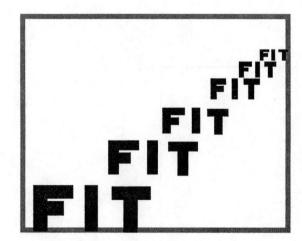

19. $d = 12.5, \ c = 7.5$

$$\frac{a}{e} = \frac{b}{d}$$

$$\frac{4}{20} = \frac{2.5}{d}$$

$$4d = 50$$

$$d = 12.5$$

$$c = e - d = 20 - 12.5 = 7.5$$

21. $b \approx 2.2, \ c \approx 20$

$$\frac{a}{e} = \frac{b}{d}$$

$$\frac{12.2}{24.5} = \frac{b}{4.5}$$

$$24.5b = 54.9$$

$$b \approx 2.2$$

$$c = e - d = 24.5 - 4.5 = 20$$

23. $d = 15, \ b = 6$

$$d = e - c = 20 - 5 = 15$$

$$\frac{a}{e} = \frac{b}{d}$$

$$\frac{8}{20} = \frac{b}{15}$$

$$20b = 120$$

$$b = 6$$

Explore

25.

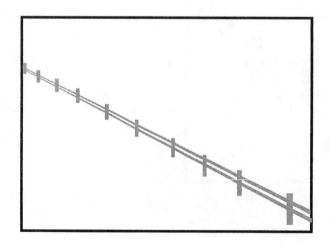

27. overlapping shapes

29. one point perspective

31. None of the methods are used. This is an example of a "flat painting."

33. 13 1/3 in.

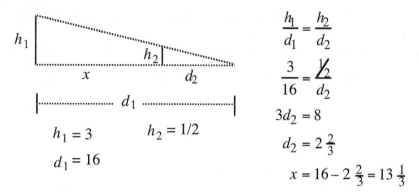

$$\frac{h_1}{d_1} = \frac{h_2}{d_2}$$

$$\frac{3}{16} = \frac{\frac{1}{2}}{d_2}$$

$$3d_2 = 8$$

$$d_2 = 2\frac{2}{3}$$

$$x = 16 - 2\frac{2}{3} = 13\frac{1}{3}$$

$h_1 = 3$ $h_2 = 1/2$

$d_1 = 16$

35. The tracks seem to get closer together. This suggests a one-point perspective.

37. Answer depends on what work of art is examined.

SECTION 7.3 GOLDEN RATIOS AND RECTANGLES

This section introduces the **Golden Ratio** (ϕ). The ancient Greeks believed that this ratio was pleasing to the eye. This ratio is based on the division of a line segment into two parts such that the ratio of the longer piece to the shorter piece is the same as the ratio of the entire segment to the longer piece. The Golden Ratio has an exact value of $\frac{1+\sqrt{5}}{2}$ or approximately 1.62.

The Golden Ratio

$$\phi = \frac{a}{b} = \frac{a+b}{a} \approx 1.62$$

A rectangle that has the ratio of its length to its width equal to the Golden Ratio is called a Golden Rectangle. The section contains various applications of and experiments with the Golden Ratio.

Explain

1. The Golden Ratio (ϕ) is a ratio between distances a and b such that,
   $$\phi = \frac{a}{b} = \frac{a+b}{a} \approx 1.62.$$

3. Many artists and sculptors use the Golden Ratio in their work because such a ratio is more pleasing to the eye.

5. A Golden Rectangle is rectangle with the ratio of its sides equal to the Golden Ratio.

7. If l = the longer side and s = the shorter side, substitute the known value for s and solve for l using the ratio $\dfrac{l}{s} \approx 1.62$.

Apply

9. The point is 2.6 cm from one end of the segment.
   The line segment is 4.2 cm long. Thus,
   $$\frac{a+b}{a} = \frac{4.2}{a} = 1.62$$
   $$1.62a = 4.2$$
   $$a = 2.6.$$

11. The point is 3.8 cm from one end of the segment.
    The line segment is 6.3 cm long. Thus,
    $$\frac{a+b}{a} = \frac{6.2}{a} = 1.62$$
    $$1.62a = 6.2$$
    $$a = 3.8$$

13. No

15. No

17. 37.3 ft, 14.2 ft

    If the shorter segment = 23 ft,
    $$\frac{a}{b} = 1.62$$
    $$\frac{a}{23} = 1.62$$
    $$a = 37.3.$$

    If the longer segment = 23 ft,
    $$\frac{a}{b} = 1.62$$
    $$\frac{23}{b} = 1.62$$
    $$1.62b = 23$$
    $$b = 14.2.$$

19. 92.0 m, 35.1 m

If the shorter segment $= 56.8$ m,

$$\frac{a}{b} = 1.62$$

$$\frac{a}{56.8} = 1.62$$

$$a = 92.0.$$

If the longer segment $= 56.8$ m,

$$\frac{a}{b} = 1.62$$

$$\frac{56.8}{b} = 1.62$$

$$1.62b = 56.8$$

$$b = 35.1.$$

21. 14.6 cm, 5.6 cm,

If the shorter segment $= 9$ cm,

$$\frac{a}{b} = 1.62$$

$$\frac{a}{9} = 1.62$$

$$a = 14.6.$$

If the longer segment $= 9$ cm,

$$\frac{a}{b} = 1.62$$

$$\frac{9}{b} = 1.62$$

$$1.62b = 9$$

$$b = 5.6.$$

23. 58.3 in., 22.2 in.

If the shorter segment $= 36$ in.,

$$\frac{a}{b} = 1.62$$

$$\frac{a}{36} = 1.62$$

$$a = 58.3.$$

If the longer segment $= 36$ in.,

$$\frac{a}{b} = 1.62$$

$$\frac{36}{b} = 1.62$$

$$1.62b = 36$$

$$b = 22.2.$$

Explore

25. $w = 9.9$ cm, $b = 9.9$ cm, $t = 6.1$ cm

$$\frac{h}{w} = 1.62 \qquad\qquad \frac{b+t}{b} = 1.62 \qquad\qquad t = h - b$$

$$\frac{16}{w} = 1.62 \qquad\qquad \frac{16}{b} = 1.62 \qquad\qquad t = 16 - 9.9 = 6.1$$

$$1.62w = 16 \qquad\qquad 1.62b = 16$$

$$w = 9.9 \qquad\qquad\quad b = 9.9$$

27. Answers vary.

29. $w = 8.1$ in., $l = 13.1$ in.

$$\frac{w}{h} = 1.62 \qquad\qquad \frac{l}{w} = 1.62$$

$$\frac{w}{5} = 1.62 \qquad\qquad \frac{l}{8.1} = 1.62$$

$$w = 8.1 \qquad\qquad\quad w = 13.1$$

31. $w = 5.6$ cm, $l = 14.6$ cm

$$\frac{w}{h} = 1.62 \qquad\qquad \frac{l}{w} = 1.62$$

$$\frac{9}{h} = 1.62 \qquad\qquad \frac{l}{9} = 1.62$$

$$1.62h = 9 \qquad\qquad w = 14.6$$

$$h = 5.6$$

33. $w = 14.6$ cm, $l = 23.7$ cm

$$\frac{w}{h} = 1.62 \qquad\qquad \frac{l}{w} = 1.62$$

$$\frac{w}{9} = 1.62 \qquad\qquad \frac{l}{14.6} = 1.62$$

$$w = 14.6 \qquad\qquad l = 23.7$$

35. Requires construction of a box with $h = 3$ in., $w = 4.9$ in., and $l = 7.9$ in.

37. The ratios of consecutive Fibonacci numbers are approximate the Golden Ratio. Thus, three consecutive Fibonacci numbers give approximate Golden boxes.

39. Requires measurement of cans.

41. Requires a survey of ten people using different parallelograms.

43.

$$A + B = 180, \text{ so } B = 180 - A$$

$$\text{Since } \frac{B}{A} = \frac{180 - A}{A} = \frac{1 + \sqrt{5}}{2},$$

$$360 - 2A = A + A\sqrt{5}$$

$$360 = 3A + A\sqrt{5}$$

$$360 = A\left(3 + \sqrt{5}\right)$$

$$\frac{360}{3 + \sqrt{5}} = A$$

A **polygon** is a closed figure in a plane formed by segments that intersect each other only at their end points. Regular polygons are those with equal sides and equal angles. A regular polygon can be drawn by finding equally spaced points on a circle and connecting consecutive points. The sum of the angles of a polygon with n-sides can be determined by the formula,

$$S = 180(n - 2).$$

If the polygon is a regular polygon, each angle (A) of the regular polygon can be determined by the formula,

$$A = \frac{180(n - 2)}{n}.$$

If a polygon has n-sides, a central angle of $360°/n$ can be used to locate the equally spaced points on the circle. To create a star, non-consecutive equally spaced points are systematically connected.

Explain

1. A polygon is a closed figure in a plane formed by line segments that intersect each other only at their end points.

3. The figure is not closed.

5. Sides cross each other.

7. Angles are not equal.

9. Sides are not equal.

11. Concave polygons are polygons in which an extension of at least one of its sides enter the interior of the polygon. In a convex polygon the extensions of the sides do not enter the interior of the polygon.

13. Concave 14-gon.

15. Convex heptagon.

17. Find n equally spaced point on a circle by using central angles of $360°/n$. Connect non-consecutive points in a systematic pattern (i.e. every third point) to get a regular star.

19. To find the each angle of a regular polygon with n sides used the formula:
$$A = \frac{180(n-2)}{n}.$$

Apply

21. $S = 540°$

A pentagon has five sides. Using $n = 5$, we get
$$S = 180(n-2)$$
$$S = 180(5-2) = 180(3) = 540°.$$

(a)    (b)

23. $S = 1980°$

A 13-gon has 13 sides. Using $n = 13$, we get
$$S = 180(n-2)$$
$$S = 180(13-2) = 180(11) = 1980°.$$

(a) 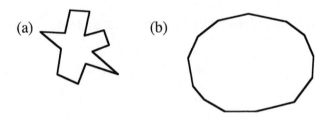   (b)

25. $S = 1440°$

A decagon has ten sides. Using $n = 10$, we get
$$S = 180(n-2)$$
$$S = 180(10-2) = 180(8) = 1440°.$$

(a) 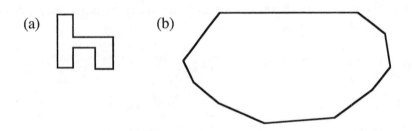   (b)

27. Since the central angle is determined by dividing 360° by the number of points desired, we get the following.

Points	Angle	Points	Angle
5	72.0°	13	$\approx 27.7°$
6	60.0°	14	$\approx 25.7°$
7	$\approx 51.4°$	15	24.0°
8	45.0°	16	22.5°
9	40.0°	18	20.0°
10	36.0°	20	18.0°
11	32.7°	30	12.0°
12	30.0°	36	10.0°

29. $A = 135°$

A regular octagon has eight sides. Using $n = 8$, we get
$$A = \frac{180(n-2)}{n}$$
$$A = \frac{180(8-2)}{8} = \frac{180(6)}{8} = 135°.$$

31. $A \approx 147.3°$

A regular undecagon has eleven sides. Using $n = 11$, we get
$$A = \frac{180(n-2)}{n}$$
$$A = \frac{180(11-2)}{11} = \frac{180(9)}{11} \approx 147.3°.$$

33. $A = 162°$

A regular 20-gon has twenty sides. Using $n = 20$, we get

$$A = \frac{180(n-2)}{n}$$

$$A = \frac{180(20-2)}{20} = \frac{180(18)}{11} = 162°.$$

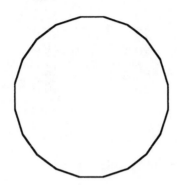

35. $A \approx 154.3°$

A regular 14-gon has twenty sides. Using $n = 14$, we get

$$A = \frac{180(n-2)}{n}$$

$$A = \frac{180(14-2)}{14} = \frac{180(12)}{14} \approx 154.3°.$$

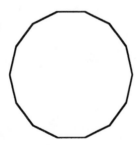

Explore

37. Mark off 10 equally spaced points on a circle using a central angle of 36° and connect every third point.

39. Mark off 15 equally spaced points on a circle using a central angle of 24° and connect every sixth point.

41. Mark off 20 equally spaced points on a circle using a central angle of 18° and connect every ninth point.

A **tessellation** is a pattern of one or more congruent shapes that cover a region of a plane without overlapping or leaving any gaps. A set of polygons will tessellate a region if the sum of the angles of the polygons located around each vertex has a sum of exactly 360°. The formula for each angle (A) of a regular polygon of n-sides, $A = 180(n - 2)/n$, can be used to help determine if certain regular polygons can be used to tessellate an area. The tessellation of a region by triangles and quadrilaterals can be accomplished using the geometric operations of translations (slides) and reflections (flips).

Explain

1. A tessellation is a pattern of one or more congruent shapes that covers an area in a plane without overlapping or leaving any gaps.

3. The three regular polygons that can tessellate a plane are the equilateral triangle, the square, and the regular hexagon.

5. A translation is the movement of a shape in a plane by sliding the shape in a certain direction for a specified distance.

7. Create a primary line of triangles by repeatedly translating the original triangle in the direction of one of the sides of the triangle. Using each triangle in that primary line repeatedly translate in the direction of another side of the triangle. Connect any missing lines.

9. Repeatedly translate the quadrilateral in the direction of both diagonals.

Apply

11. A regular heptagon has angles of 128 4/7°. Since 360° is not evenly divisible by 128 4/7°, regular heptagons will not tessellate the plane.

13. A regular decagon has angles of 144°. Since 360° is not evenly divisible by 144°, a regular decagon will not tessellate the plane.

In the answers to Problems 14 – 25, each diagram is approximately 40% of actual size and shows four to seven tiles.

15.

17.

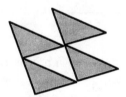

19.

21.

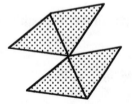

23.

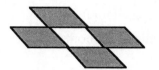

25.

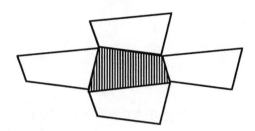

Explore

27. The sum of the angles for these three polygons does not add to 360°.

    Each angle for a regular dodecagon is 150° and each angle for an octagon is 135°.
    Since 150° + 150° + 135° = 435°, overlapping will occur and it will be impossible to
    tessellate a region with those polygons.

29.

31. Two possible answers are to use two octagons and a square or two pentagons and a
    decagon.

33. Answers vary.

35.

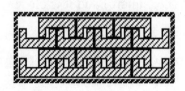

37. Let A = each angle of a regular polygon with n sides.

Since the sum of the angles around each vertex in a tessellation must equal $360°$, A must divide evenly into $360°$. The expression $\dfrac{360}{A}$ must be a nonzero whole number.

$$\frac{360}{A} = \frac{360}{\frac{180(n-2)}{n}}$$

$$= 360 \times \frac{n}{180(n-2)}$$

$$= \frac{2n}{n-2}$$

Only $n = 3$, 4, and 6 will make this expression a nonzero whole number.

SECTION 7.6 FRACTALS

This section gives a short history of fractals, describes how fractals can be drawn, gives examples of famous fractal sets, and explains fractal dimensions. The dimension of a fractal can be determined by the formula given below.

$$d = \frac{\log N}{\log \frac{1}{r}} \quad \text{where} \quad \begin{cases} r = \text{ratio of the length of the new object} \\ \quad \text{to the length of the original object} \\ N = \text{the number of new objects} \end{cases}$$

The phenomenon of self-similarity can be seen in all fractals. The same patterns reappear as the fractal is examined more and more closely. The use of computational and graphics capabilities of modern computers have allowed us to visualize the world of fractal geometry.

Explain

1. A fractal is a object with a fractional dimension. Fractals have the quality of self-similarity, that is, it contains repeated smaller versions of the same pattern.

3. A dimension of 1.6 indicates that the object is between a one-dimensional object (i.e. line segment) and a two-dimensional object (i.e. square) and it is closer to a two-dimensional object.

5. Self-similarity means that the object contains repeated smaller versions of the same pattern.

7. The leaf of a fern is like a fractal since each leaflet is similar to the entire leaf of the fern and each sub-leaflet is similar to the leaflet. It possesses self-similarity.

Apply

9.

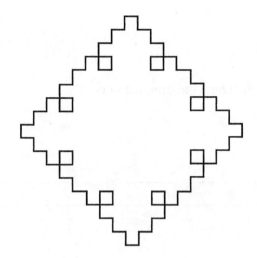

11.

13.

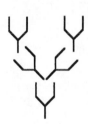

15. Answers vary.

Explore

17. 1.46

In this fractal, $r = 1/3$ and $N = 5$. Thus, the dimension is

$$d = \frac{\log 5}{\log\left(\frac{1}{1/3}\right)} = \frac{\log 5}{\log 3} \approx 1.46.$$

19. 1.29

In this fractal, $r = 1/4$ and $N = 6$. Thus, the dimension is

$$d = \frac{\log 6}{\log\left(\frac{1}{1/4}\right)} = \frac{\log 6}{\log 4} \approx 1.29.$$

21. 1.46

In this fractal, $r = 1/3$ and $N = 5$. Thus, the dimension is

$$d = \frac{\log 5}{\log\left(\frac{1}{1/3}\right)} = \frac{\log 5}{\log 3} \approx 1.46.$$

23. 0.68

In this fractal, $r = 1/5$ and $N = 3$. Thus, the dimension is

$$d = \frac{\log 3}{\log\left(\frac{1}{1/5}\right)} = \frac{\log 3}{\log 5} \approx 0.68.$$

CHAPTER 7 REVIEW

Review Section 7.1

1. It is necessary to have undefined terms and postulates in a deductive system of geometry because there must be something to start the system. Not all terms can be defined nor can all propositions be deduced.

2.
$$6x + 5 = 2x - 3$$
$$6x - 2x + 5 = 2x - 2x - 3 \quad \text{(A3)}$$
$$4x + 5 = -3$$
$$4x + 5 - 5 = -3 - 5 \quad \text{(A3)}$$
$$4x = -8$$
$$x = -2 \quad \text{(A4)}$$

3. Euclidean

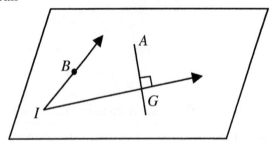

Lobachevskian

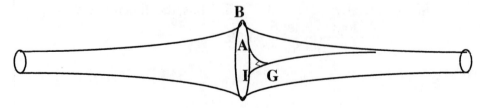

Riemannian

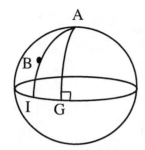

4.   Euclidean

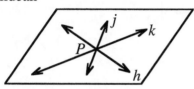

Lobachevskian

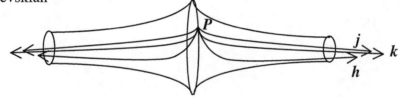

Riemannian

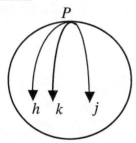

5.   A square has an angle sum of 360° (four right angles). However, in Lobachevskian geometry, the angle sum of a quadrilateral is less than 360° and in Riemannian geometry, the sum is greater than 360°. Thus, squares do not exist.

Review Section 7.2

6.   The artist uses overlapping shapes, diminishing sizes, and atmospheric perspective in a one-point perspective to create the inside of the building.

7.   None of the techniques used in representing a three-dimensional scene on a two-dimensional canvas are used in the wall painting.

8.

9. 1.8 in.

x = the height of the cross

$$\frac{6}{10} = \frac{x}{3}$$

$$10x = 18$$

$$x = 1.8$$

Review Section 7.3

10. The Greeks believed that if the ratio of distances equaled the Golden Ratio, it was pleasing to the eye. The Golden Ratio was used in Greek art and construction.

11. 13.6 ft. from one end of the stage

$$\frac{22}{x} = 1.62$$

$$1.62x = 22$$

$$x = 13.6$$

12. 87.5", 33.3"

x = longer side

$$\frac{x}{54} = 1.62$$

$$x = 1.62(54)$$

$$x = 87.5$$

x = shorter side

$$\frac{54}{x} = 1.62$$

$$1.62x = 54$$

$$x = 33.3$$

13. Either the ratio of the diagonals or the ratio of the sides should be 1.62.

Review Section 7.4

14. A dodecagon has 12 sides. Use 12 equally spaced points on circle using a central angle of 30° for the regular dodecagon.

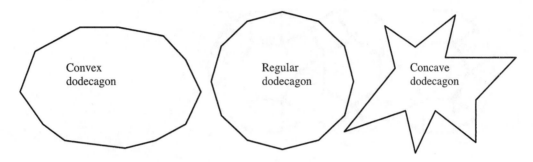

15. 1800°

$$S = 180(n - 2)$$
$$= 180(12 - 2)$$
$$= 1800$$

16. 150°

$$A = \frac{180(n - 2)}{n}$$
$$= \frac{1800}{12}$$
$$= 150$$

17.

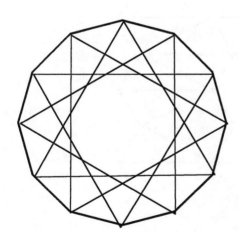

Review Section 7.5

18. From Problem 16, each angle of a regular dodecagon is 150°. No combination of 150° angles will add up to 360°. Therefore, regular dodecagons can not tessellate a region.

19.

20. (a)

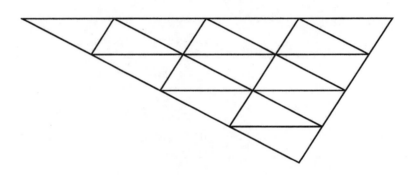

(b)

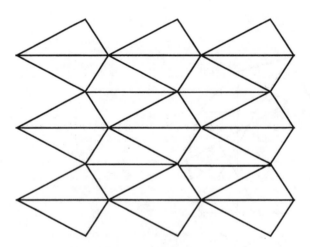

21. (a)

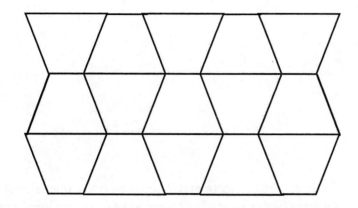

(b)

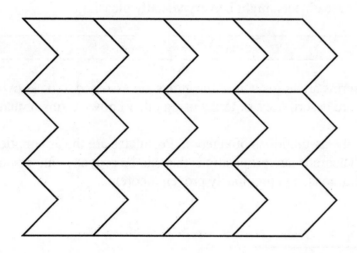

Review Section 7.6

22.

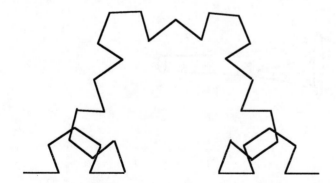

23. 1.77

$$N = 7, \quad r = \tfrac{1}{3}$$

$$d = \frac{\log N}{\log(1/r)}$$

$$= \frac{\log 7}{\log(3)}$$

$$= 1.77$$

24. A dimension of 1.62 indicates that objects exists between one-dimensional and two-dimensional space. Objects in this space are closer to two-dimensional objects than one-dimensional objects. This dimension would be "special," since 1.62 is the Golden Ratio. These objects might be very visually pleasing.

CHAPTER 7 TEST

1. (a) Undefined terms are terms whose meanings are assumed to be known. Definitions are descriptions of terms using other known terms or undefined terms.
    (b) Axioms are the assumptions of algebra. Postulates are the assumptions of geometry. Theorems are propositions that can be proven using postulates, axioms, definitions, and previously proven theorems.

2. Euclidean

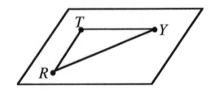

Lobachevskian

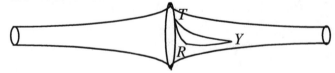

Riemannian

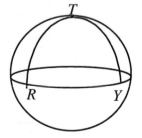

3. Mathematics does give truth about the real world. It can only give logical descriptions of the real world based on beginning assumptions (postulates).

4. (a)

(b)

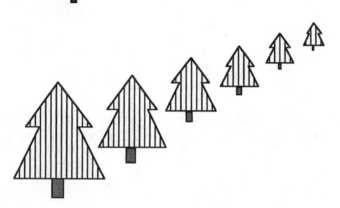

(c)

5.

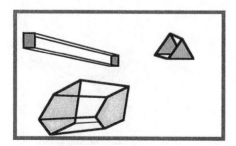

6. The point is 1.1 in. from one end of the segment.
    The line segment is 1.75 in. long. Thus,

    $$\frac{a+b}{a} = \frac{1.75}{a} = 1.62$$
    $$1.62a = 1.75$$
    $$a \approx 1.1.$$

7. 13.8 inches
    $$\frac{l}{w} = 1.62$$
    $$\frac{l}{8.5} = 1.62$$
    $$l \approx 13.8$$

8. (a) Use 12 equally spaced points on
         circle and a central angle of 30°.

    (b) Use 18 equally spaced points on
         circle and a central angle of 20°.

9. (a)

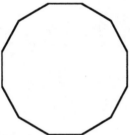

    (b)

10. (a)

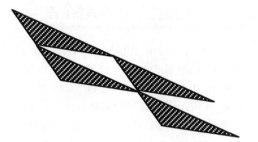

(b)

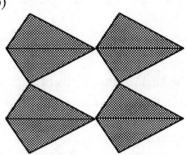

11. (a)

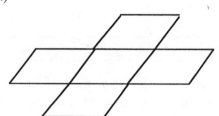

(b)

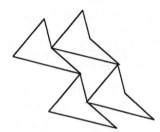

12. For a regular polygon to tessellate a region, the sum of the angles around any vertex must equal 360°. The angles of a square (90°), a regular hexagon (120°), and a regular dodecagon (150°) have a sum of 360°. Thus, they can tessellate a plane. The angles of an equilateral triangle (60°), a regular pentagon (108°), and a regular heptagon (128 4/7°) have a sum of 296 4/7°. Thus, they cannot tessellate a plane.

13.

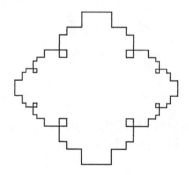

14. 1.29

In this fractal, $r = 1/4$ and $N = 6$. Thus, the dimension is:

$$d = \frac{\log 6}{\log\left(\frac{1}{1/4}\right)} = \frac{\log 6}{\log 4} \approx 1.29.$$

CHAPTER 8   TRIGONOMETRY: A DOOR TO THE UNMEASURABLE

In this chapter, you will learn how to find the unknown sides and angles of triangles by using some of the basic components of trigonometry. With the use of trigonometric formulas and a scientific calculator, you will be able to determine distances that you may not be able to measure directly. For example, you will be able to find the height of a building, the width of a canyon, the altitude of a hot-air-balloon, the distance to a forest fire or the distance of a shotput. You will see that mathematics can be used to solve some very practical problems. You will see that a knowledge of trigonometry opens a door to determining unmeasurable distances.

SECTION 8.1   RIGHT TRIANGLES, SINE, COSINE, TANGENT

This section examines the difference between the legs and the hypotenuse of a right triangle and the difference between an adjacent side and an opposite side for an acute angle of the right triangle. This section also examines the Pythagorean Theorem. This theorem states that for a right triangle, if a and b are the lengths of the legs of the right triangle and c is the length of the hypotenuse, then, $a^2 + b^2 = c^2$. The section introduces the trig functions – sine, cosine, tangent, and the acronym, **soh cah toa**, that can be used to remember the trig ratios.

$$\sin A = \frac{\text{opposite}}{\text{hypotenuse}} = \frac{a}{c}$$

$$\cos A = \frac{\text{adjacent}}{\text{hypotenuse}} = \frac{b}{c}$$

$$\tan A = \frac{\text{opposite}}{\text{adjacent}} = \frac{a}{b}$$

The section also shows how to use a scientific calculator to find the ratio of sides for a given angle and how to find an angle from a ratio of sides in a right triangle.

Explain

1. A right triangle is a triangle with a 90° angle.

3. The Pythagorean Theorem is the relationship involving the sides of a right triangle which states that, if a and b are legs and c is the hypotenuse, $a^2 + b^2 = c^2$.

5. The sine of an acute angle of a right triangle is the ratio of the length of the side opposite that angle to the length of the hypotenuse. The cosine of an angle of a right triangle is the ratio of the length of the side adjacent to that angle to the length of the

hypotenuse. The tangent of an acute angle of a right triangle is the ratio of the length of the side opposite that angle to the length of the side adjacent to that angle.

7. *M* represents an acute angle of the right triangle, *k* represents the side adjacent to ∠*M*, and *f* represents the hypotenuse of the triangle.

9. The chant-like acronym, *soh cah toa*, gives a way to remember the definitions of the trig functions: $soh - \sin(angle) = \dfrac{\text{opposite}}{\text{hypotenuse}}$, $cah - \cos(angle) = \dfrac{\text{adjacent}}{\text{hypotenuse}}$, $toa - \tan(angle) = \dfrac{\text{opposite}}{\text{adjacent}}$.

Apply

11. legs: *k*, *d*; hypotenuse: *y*

13. (10) opposite: y; adjacent: *d*
    (11) opposite: *d*; adjacent: *k*
    (12) opposite: *k*; adjacent: *y*

15. 10.6

$$t^2 = n^2 + x^2$$
$$t^2 = 9.7^2 + 4.3^2$$
$$t^2 = 94.09 + 18.49$$
$$t^2 = 112.58$$
$$t = 10.6$$

17. 67.2

$$t^2 = n^2 + x^2$$
$$67.24^2 = 3.24^2 + x^2$$
$$4521.2176 = 10.4976 + x^2$$
$$4510.72 = x^2$$
$$67.2 = x$$

19. $\sin A = 3/5$ $\sin B = 4/5$
    $\cos A = 4/5$ $\cos B = 3/5$
    $\tan A = 3/4$ $\tan B = 4/3$

21. $\sin A = 4/5$ $\sin B = 3/5$
     $\cos A = 3/5$ $\cos B = 4/5$
     $\tan A = 4/3$ $\tan B = 3/4$

23. 0.2181

25. 0.9947

27. 4.1022

29. 82.9°

31. 24.6°

33. 73.7°

Explore

35. yes, $\left(36^2 + 48^2 = 60^2\right)$

37. yes, $\left(33^2 + 56^2 = 65^2\right)$

39. yes, $\left(5^2 + 12^2 = 13^2\right)$

41. 36.9°, 53.1°

$$\tan A = \frac{3}{4} = 0.75$$
$$A = \tan^{-1}(0.75) = 36.9°$$
$$B = 90 - 36.9° = 53.1°$$

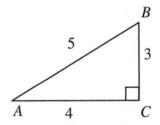

43. 43.6°, 46.4°

$$\sin A = \frac{20}{29} = 0.6897$$
$$A = \sin^{-1}(0.6897) = 43.6°$$
$$B = 90 - 43.6° = 46.4°$$

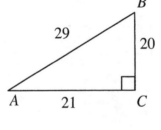

45. 71.1°, 18.9°

$$\cos B = \frac{35}{37} = 0.9459$$
$$B = \cos^{-1}(0.9459) = 18.9°$$
$$A = 90 - 18.9° = 71.1°$$

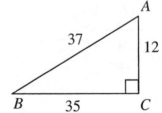

47. $AC = 10$, $AD = 12.2$, $\angle DAC = 35.0°$

$$AC^2 = 6^2 + 8^2 \qquad AD^2 = 10^2 + 7^2$$
$$AC^2 = 36 + 64 \qquad AD^2 = 100 + 49$$
$$AC^2 = 100 \qquad AD^2 = 149$$
$$AC = 10 \qquad AD = 12.2$$

$$\tan \angle DAC = \frac{7}{10} = 0.7$$
$$\angle DAC = \tan^{-1}(0.7)$$
$$= 35.0°$$

49. $AC = 7.6$, $AD = 9.2$, $\angle DAC = 34.4°$

$$AC^2 = 3.9^2 + 6.5^2 \qquad AD^2 = 5.2^2 + 7.6^2$$
$$AC^2 = 15.21 + 42.25 \qquad AD^2 = 27.04 + 57.76$$
$$AC^2 = 57.46 \qquad AD^2 = 84.8$$
$$AC = 7.6 \qquad AD = 9.2$$

$$\tan \angle DAC = \frac{5.2}{7.6} = 0.6842$$
$$\angle DAC = \tan^{-1}(0.6842)$$
$$= 34.4°$$

SECTION 8.2   Solving Right Triangles

This section shows how to solve right triangles, that is, to find unknown sides and angles of right triangles. Typically, three parts of a right triangle (sides or angles) are known. You will use the trig functions – sine, cosine, tangent, the Pythagorean Theorem, and the fact that the sum of the angles of a triangle equals 180° to find the other three parts of the triangle. To establish uniformity of answers the following round-off rules will be used in the rest of the chapter.

> **Round-Off Rules**
>
> The **final answers** to angles and sides of triangles that have more than one decimal digit will be rounded off to the **nearest tenth**. The final answers will be computed using intermediate results that have been rounded off to four decimal places.

Explain

1. You should find the lengths of the sides and measure of the angles of the triangle.

3. The final answers to the sides of triangles that have more than one decimal digit should be rounded off to the nearest tenth.

5. Use the tangent ratio of the two known legs, to find one of the acute angles. Subtract the measure of that angle from 90° to find the other acute angle.

7. You could (a) use the tangent ratio of the two known legs, to find one of the acute angles (b) use the sine ratio of the known leg and the hypotenuse, to find one of the acute angles, or (c) use the cosine ratio of the other leg and the hypotenuse, to find one of the acute angles. Subtract the measure of that angle from 90° to find the other acute angle.

Apply

9. $\angle C = 53.1°$, $\angle A = 36.9°$, $b = 6.5$

$\tan C = \dfrac{5.2}{3.9}$

$\tan C = 1.3333$

$C = 53.1°$

$\angle A + \angle C = 90°$

$\angle A + 53.1° = 90°$

$\angle A = 36.9°$

$\cos 53.1° = \dfrac{3.9}{b}$

$0.6004 = \dfrac{3.9}{b}$

$0.6004\,b = 3.9$

$b = 6.5$

11. $\angle A = 41°$, $b = 13.6$, $a = 11.8$

$\angle A + \angle B = 90°$

$\angle A + 49° = 90°$

$\angle A = 41°$

$\sin 49° = \dfrac{b}{18}$

$0.7547 = \dfrac{b}{18}$

$13.6 = b$

$c^2 = a^2 + b^2$

$18^2 = a^2 + 13.6^2$

$324 = a^2 + 184.96$

$139.04 = a^2$

$11.8 = a$

13. $\angle X = 73.3°$, $y = 39.0$, $z = 135.7$

$\angle X + \angle Y = 90°$

$\angle X + 16.7° = 90°$

$\angle X = 73.3°$

$\tan 16.7° = \dfrac{y}{130}$

$0.3000 = \dfrac{y}{130}$

$39.0 = y$

$\sin 16.7° = \dfrac{39.0}{z}$

$0.2874 = \dfrac{39.0}{z}$

$0.2874\,z = 39.0$

$z = 135.7$

15. $\angle X = 75.8°$, $\angle Y = 14.2°$, $z = 1004.6$

$\tan X = \dfrac{974}{246}$

$\tan X = 3.9593$

$X = 75.8°$

$Y = 90 - 75.8° = 14.2°$

$z^2 = x^2 + y^2$

$z^2 = 974^2 + 246^2$

$z^2 = 1,009,19$

$z = 1004.6$

Explore

17. $\angle T = 2.2°, \angle P = 87.8°, p = 2600$

$$\sin T = \frac{102}{2602}$$

$$\sin T = 0.0392$$

$$T = 2.2°$$

$$P = 90° - 2.2° = 87.8°$$

$$i^2 = p^2 + t^2$$

$$2602^2 = p^2 + 102^2$$

$$6,770,404 = p^2 + 10,404$$

$$6,760,00 = p^2$$

$$2600 = p$$

19. $\angle T = 74.6°, i = 783.1, t = 755.3$

$$T = 90° - 15.4° = 74.6°$$

$$\sin 15.4 = \frac{208}{i}$$

$$0.2656 = \frac{208}{i}$$

$$0.2656i = 208$$

$$i = 783.1$$

$$\tan 15.4 = \frac{208}{t}$$

$$0.2754 = \frac{208}{t}$$

$$0.2754t = 208$$

$$t = 755.3$$

21. $\angle P = 67°, i = 48.5, t = 18.9$

$$T = 90° - 23° = 67°$$

$$\tan 23 = \frac{t}{44.6}$$

$$0.4248 = \frac{t}{44.6}$$

$$18.9 = t$$

$$\cos 23 = \frac{44.6}{i}$$

$$0.9205 = \frac{44.6}{i}$$

$$0.9205i = 44.6$$

$$i = 48.5$$

23. $h = 35.4$

From the smaller triangle, we have

$$\tan 45° = \frac{h}{x}$$

$$1 = \frac{h}{x}$$

$$x = h.$$

From the larger triangle, we have

$$\tan 23° = \frac{h}{x + 48}$$

$$0.4245 = \frac{h}{x + 48}$$

$$0.4245(x + 48) = h$$

$$0.4245x + 20.376 = h.$$

Substituting $x = h$ gives
$$0.4245h + 20.376 = h$$

$$20.376 = 0.5755h$$

$$35.4 = h.$$

25. $h = 92.4$

From the smaller triangle, we have

$$\tan 57° = \frac{h}{x}$$

$$1.5399 = \frac{h}{x}$$

$$1.5399x = h$$

$$x = \frac{h}{1.5399}.$$

From the larger triangle, we have

$$\tan 30° = \frac{h}{x + 100}$$

$$0.5774 = \frac{h}{x + 100}$$

$$0.5774(x + 100) = h$$

$$0.5774x + 57.74 = h.$$

Substituting $x = \frac{h}{1.5399}$ gives

$$0.5774\left(\frac{h}{1.5399}\right) + 57.74 = h$$

$$0.3750h + 57.74 = h$$

$$57.74 = 0.6250h$$

$$92.4 = h.$$

27. $x = 14.5$

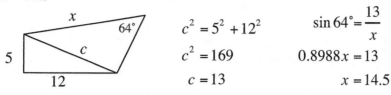

$$c^2 = 5^2 + 12^2$$

$$c^2 = 169$$

$$c = 13$$

$$\sin 64° = \frac{13}{x}$$

$$0.8988x = 13$$

$$x = 14.5$$

SECTION 8.3   RIGHT TRIANGLE APPLICATIONS

The three trigonometric functions can be applied to various situations involving right triangles. It will be helpful to:

1. Sketch a triangle that fits the situation.
2. Write the values of the known angles and sides on the triangle.
3. Use letters to represent the unknown angles and sides.
4. Use trig functions to solve for an unknown in a trig equation.
5. Answer the question posed in the problem using the round off rules (Section 8.2).

Explain

1. (a) Make a sketch of the situation and the right triangle involved.
    (b) Write the known angles and sides on the triangle.
    (c) Use letters to represent the unknown angles and sides.
    (d) Use trigonometric functions to help solve the triangle.

3. An angle of depression is an angle made between a horizontal line and a line to an object that is below the horizontal line.

5. The use of trigonometry allows you to determine sides and angles of right triangles that may be very difficult or even impossible to physically measure. Thus, the title, "Trigonometry: A Door to the Unmeasureable," seems appropriate.

Apply

7. 2.6 mi

$$\tan A = \frac{1350}{y}$$

$$0.10 = \frac{1350}{y}$$

$$0.10y = 1350$$

$$y = 13,500$$

$$x^2 = 13,500^2 + 1350^2$$

$$x^2 = 184,072,500$$

$$x = 13,567.3 \text{ ft}$$

$$x = 13,567.3/5280 \approx 2.6 \text{ mi}$$

9.　231.8 m

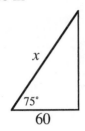

$$\cos 75° = \frac{60}{x}$$

$$0.2588 = \frac{60}{x}$$

$$0.2588x = 60$$

$$x = 231.8$$

Explore

11.　pole: 50.0 ft, wire: 51.4 ft

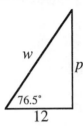

$$\tan 76.5° = \frac{p}{12}$$

$$4.1653 = \frac{p}{12}$$

$$50.0 = p$$

$$\cos 76.5° = \frac{12}{w}$$

$$0.2334 = \frac{12}{w}$$

$$0.2334w = 12$$

$$w = 51.4$$

13.　29.5 ft

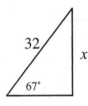

$$\sin 67° = \frac{x}{32}$$

$$0.9211 = \frac{x}{32}$$

$$29.5 = x$$

15.　114.9 ft

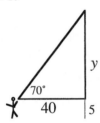

$$\tan 70° = \frac{y}{40}$$

$$2.7475 = \frac{y}{40}$$

$$109.9 = y$$

$$\text{height} = y + 5 = 114.9 \text{ ft}$$

17. 2192.9 ft

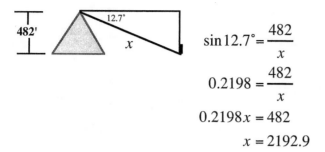

$$\sin 12.7° = \frac{482}{x}$$

$$0.2198 = \frac{482}{x}$$

$$0.2198x = 482$$

$$x = 2192.9$$

19. 65.3 sec

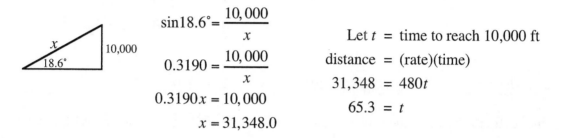

$$\sin 18.6° = \frac{10,000}{x}$$

$$0.3190 = \frac{10,000}{x}$$

$$0.3190x = 10,000$$

$$x = 31,348.0$$

Let t = time to reach 10,000 ft

distance = (rate)(time)

$$31,348 = 480t$$

$$65.3 = t$$

21. 14.1 ft, 12.1°

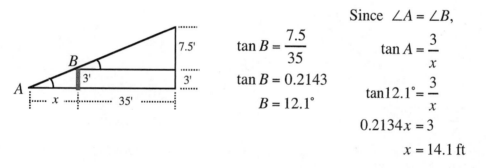

$$\tan B = \frac{7.5}{35}$$

$$\tan B = 0.2143$$

$$B = 12.1°$$

Since $\angle A = \angle B$,

$$\tan A = \frac{3}{x}$$

$$\tan 12.1° = \frac{3}{x}$$

$$0.2134x = 3$$

$$x = 14.1 \text{ ft}$$

23. 29.5 ft

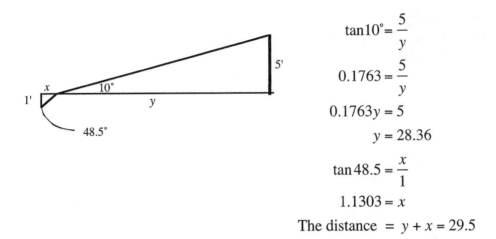

$$\tan 10° = \frac{5}{y}$$

$$0.1763 = \frac{5}{y}$$

$$0.1763y = 5$$

$$y = 28.36$$

$$\tan 48.5 = \frac{x}{1}$$

$$1.1303 = x$$

The distance $= y + x = 29.5$

This section introduces the Law of Sines and the Law of Cosines as a means of solving acute triangles.

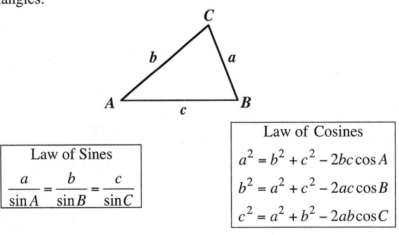

Law of Sines
$\dfrac{a}{\sin A} = \dfrac{b}{\sin B} = \dfrac{c}{\sin C}$

Law of Cosines
$a^2 = b^2 + c^2 - 2bc \cos A$
$b^2 = a^2 + c^2 - 2ac \cos B$
$c^2 = a^2 + b^2 - 2ab \cos C$

The Law of Sines can be used to solve any triangle in which two angles and one side are known. These are referred to as the angle-angle-side (**AAS**) or the angle-side-angle (**ASA**) cases. The Law of Cosines can be used to solve any triangle in which two sides and the angle between the sides (**SAS**) or the three sides (**SSS**) are known.

The case in which you are given the lengths of two sides and an angle which is not between those sides are known (**SSA**) will not be considered in this chapter since it can lead to more than one solution. This situation is referred to as the *ambiguous case* in most trigonometry books.

1. The Law of Sines states that in $\triangle ABC$, $\dfrac{a}{\sin A} = \dfrac{b}{\sin B} = \dfrac{c}{\sin C}$.

3. Use the Laws of Sines and Cosines when solving a non-right triangle.

5. The Law of Cosines should be used when two sides and the included angle (SAS) or three sides and no angles (SSS) are known.

Apply

7. $n = 26.1$

$N = 180° - 56.5° - 62.9°$ $\dfrac{n}{\sin 60.6°} = \dfrac{25}{\sin 56.5°}$

$N = 60.6°$

$$\dfrac{n}{0.8712} = 29.9801$$

$$n = 26.1$$

9. $u = 407.5$

$$N = 180° - 47° - 60°$$

$$N = 73°$$

$$\dfrac{u}{\sin 60°} = \dfrac{450}{\sin 73°}$$

$$\dfrac{u}{0.8712} = 470.5613$$

$$u = 407.5$$

11. $u = 41.0$

$$U = 180° - 66.6° - 55.5°$$

$$N = 57.9°$$

$$\dfrac{u}{\sin 57.9°} = \dfrac{44.4}{\sin 66.6°}$$

$$\dfrac{u}{0.8712} = 48.3789$$

$$u = 41.0$$

13. $n = 272.3$

$$\frac{n}{\sin 12.5°} = \frac{1240}{\sin 80.25°}$$

$$\frac{n}{0.2164} = 1258.1730$$

$$n = 272.3$$

15. $d = 32.6$

$$d^2 = 12^2 + 34^2 - 2(12)(34)\cos 73°$$

$$d^2 = 144 + 1156 - 238.5753$$

$$d^2 = 1061.4247$$

$$d = 32.6$$

17. $g = 4.1$

$$g^2 = 5.5^2 + 6.75^2 - 2(5.5)(6.75)\cos 37.25°$$

$$g^2 = 30.25 + 45.5625 - 59.1031$$

$$g^2 = 16.7094$$

$$g = 4.1$$

19. $\angle O = 60.9°$

$$8.3^2 = 9.4^2 + 5.8^2 - 2(9.4)(5.8)\cos O$$

$$68.89 = 88.36 + 33.64 - 109.04\cos O$$

$$-53.11 = -109.04\cos O$$

$$60.9° = O$$

21. $\angle C = 78.6°$, $t = 23.0$, $a = 27.4$

$$C = 180° - 56.7° - 44.7°$$

$$C = 78.6°$$

$$\frac{t}{\sin 44.7°} = \frac{32.1}{\sin 78.6°}$$

$$t = 23.0$$

$$\frac{a}{\sin 56.7°} = \frac{32.1}{\sin 78.6°}$$

$$a = 27.4$$

23. $\angle C = 66.2°$, $\angle U = 41.7°$, $\angle P = 72.1°$

$$19.8^2 = 20.6^2 + 14.4^2 - 2(20.6)(14.4)\cos C$$

$$392.04 = 424.36 + 207.36 - 593.28\cos C$$

$$-239.68 = -593.28\cos C$$

$$66.2° = C$$

$$14.4^2 = 19.8^2 + 20.6^2 - 2(19.8)(20.6)\cos U$$

$$207.36 = 392.04 + 424.36 - 815.76\cos U$$

$$-609.04 = -815.76\cos U$$

$$41.7° = U$$

$$P = 180° - 41.7° - 66.2° = 72.1°$$

25. $e = 42.3$, $\angle G = 30.9°$, $\angle T = 81.6°$

$$e^2 = 45.3^2 + 23.5^2 - 2(45.3)(23.5)\cos 67.5°$$

$$e^2 = 2052.09 + 552.25 - 814.7713$$

$$e^2 = 1789.5687$$

$$e = 42.3$$

$$23.5^2 = 45.3^2 + 42.3^2 - 2(45.3)(42.3)\cos G°$$

$$552.25 = 2052.09 + 1789.29 - 3832.38\cos G°$$

$$-3289.13 = -3832.38\cos G°$$

$$0.8582 = \cos G$$

$$30.9 = G$$

$$T = 180° - 67.5° - 30.9° = 81.6°$$

27. $\angle P = 44.9°$, $\angle E = 54.2°$, $\angle T = 80.9°$

$$20^2 = 28^2 + 23^2 - 2(23)(28)\cos P$$
$$400 = 784 + 529 - 1288\cos P$$
$$-913 = -1288\cos P$$
$$0.7089 = \cos P$$
$$44.9 = P$$

$$23^2 = 28^2 + 20^2 - 2(28)(20)\cos E$$
$$529 = 784 + 400 - 1120\cos E$$
$$-655 = -1120\cos E$$
$$0.5848 = \cos E$$
$$54.2 = E$$
$$T = 180° - 44.9° - 54.2° = 80.9°$$

29. The Law of Sines works on the right triangle $\triangle RAD$. Using either the Law of Sines or right triangle trigonometry, $AD = 42.9$ and $RD = 47.3$.

31. It is impossible to have a triangle in which the sum of the lengths of two sides is less than the length of the third side.

In this section, the Laws of Sines and Cosines are applied to many real life situations. It will be helpful if the following steps are followed when solving a problem.

1. Sketch the situation and the acute triangle involved.
2. Write the values for known angles and sides on the triangle.
3. Use letters to represent the unknown angles and sides.
4. Use the appropriate Law (Sines or Cosines) to solve the triangle. Remember to use only one unknown quantity in the trig equation.
5. Answer the questions posed in the problem using the round off rules established in Section 8.2.

Explain

1. (a) Make a sketch of the situation and the acute triangle involved.
    (b) Write the known angles and sides on the triangle.
    (c) Use letters to represent the unknown angles and sides.
    (d) If two angles are known, find the unknown angle by subtracting the sum of the two angles from 180°.
    (e) Use the Law of Sines when one side and two angles are known.
    (f) Use the Law of Cosines when only one angle or no angles are known.

3. N22°E indicates that you make an angle from the North line 22° toward the East.

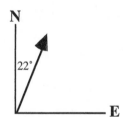

5. The use of trigonometry allows you to determine the sides and angles of acute triangles that may be very difficult or even impossible to physically measure. Thus, the title, "Trigonometry: A Door to the Unmeasureable," seems appropriate.

Apply

7. 460.0 ft

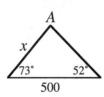

$$A = 180° - 73° - 52° = 55°$$

$$\frac{x}{\sin 52°} = \frac{500}{\sin 55°}$$

$$x = 481.0$$

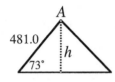

$$\sin 73° = \frac{h}{481.0}$$

$$0.9563 = \frac{h}{481.0}$$

$$460.0 = h$$

9. 27.2 ft

$$\frac{x}{\sin 80°} = \frac{25}{\sin 65°}$$

$$0.9063 x = 24.6202$$

$$x = 27.2$$

11. 33.0 ft

$$A = 180° - 63° - 63° = 54°$$

$$\frac{30}{\sin 54°} = \frac{x}{\sin 63°}$$

$$0.8090 x = 26.7302$$

$$x = 33.0$$

Explore

13. 1603.5 m

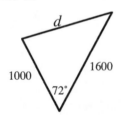

$$d^2 = 1600^2 + 1000^2 - 2(1600)(1000)\cos 72°$$

$$d^2 = 2,560,000 + 1,000,000 - 988,854.3821$$

$$d^2 = 2,571,145.6199$$

$$d = 1603.5 \text{ m}$$

15. 438.6 mi

The first plan travels for 2.5 hours at 280 mph. The distance traveled is
$d = rt = 280(2.5) = 700$ mi. The second plane travels for 2 hours at 375 mph.
The distance traveled is $d = rt = 375(2) = 750$ mi.

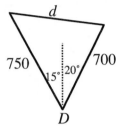

$\angle D = 15° + 20° = 35°$

$$d^2 = 750^2 + 700^2 - 2(750)(700)\cos 35°$$
$$d^2 = 562,500 + 490,000 - 860,109.6465$$
$$d^2 = 192,390.3535$$
$$d = 438.6\,\text{mi}$$

17. 12.0 ft

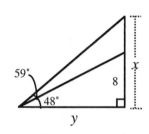

Using the bottom triangle, $\tan 48° = \dfrac{8}{y}$

$$1.1106 = \dfrac{8}{y}$$
$$1.1106y = 8$$
$$y = 7.2033.$$

Using the large triangle, $\tan 59° = \dfrac{x}{y}$

$$1.6643 = \dfrac{x}{7.2033}$$
$$12.0 = x.$$

19. 49.6°, 44.5°, 85.9°

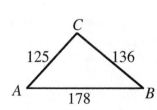

$$136^2 = 125^2 + 178^2 - 2(125)(178)\cos A$$
$$18,496 = 15,625 + 31,684 - 44,500\cos A$$
$$-28,813 = -44,500\cos A$$
$$0.6475 = \cos A$$
$$49.6° = A$$
$$125^2 = 136^2 + 178^2 - 2(136)(178)\cos B$$
$$15,625 = 18,496 + 31,684 - 48,416\cos B$$
$$-34,555 = -48,416\cos B$$
$$0.7137 = \cos B$$
$$44.5° = B$$

$$C = 180 - 49.6° - 44.5° = 85.9°$$

21. Methods may vary.

23. 1494.9 ft

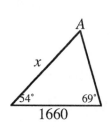

$$A = 180 - 54° - 69° = 57°$$

$$\frac{x}{\sin 69°} = \frac{1660}{\sin 57°}$$

$$\frac{x}{0.9336} = \frac{1660}{0.8387}$$

$$0.8387x = 1549.776$$

$$x = 1847.8312$$

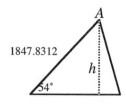

$$\sin 54° = \frac{h}{1847.8312}$$

$$0.8090 = \frac{h}{1847.8312}$$

$$1494.9 = h$$

SECTION 8.6   THE MOTION OF A PROJECTILE

In this section, we consider the motion of a projectile where gravity is the only force acting on it once it has been launched, the following set of equations models this motion:

$$x = (v \cos A)t$$
$$y = -16t^2 + (v \sin A)t + h$$

where

$$\begin{cases} t = \text{time in seconds} \\ A = \text{angle of elevation} \\ v = \text{launch velocity in } \frac{\text{ft}}{\text{sec}} \\ h = \text{height at launch in feet} \end{cases}$$

By assuming that gravity is the only force acting on the projectile once it is launched, we are assuming that the projectile is traveling in a vacuum where other factors such as air resistance, spin, drag, and friction are not considered. However, the equations are reasonably accurate in the real world if the projectile is fairly heavy and travels close to the earth at a relatively slow speed. Thus, the examples in the section focuses on the flight of motorcycles, cannonballs, shotputs, and hammer throws rather than those of golf balls, bullets, and arrows.

Explain

1. A projectile is an object that is shot forward such as a cannonball, shotput, golf ball, or bullet. The assumption is that gravity is the only force acting on the object once it has been launched.

3. The letters x, y, t, A, v, h:
$$\begin{cases} x = \text{ horizontal distance in feet} \\ y = \text{ vertical distance in feet} \\ t = \text{time in seconds} \\ A = \text{ angle of elevation in degrees} \\ v = \text{ launch velocity in ft/sec} \\ h = \text{ height at launch in feet} \end{cases}$$

5. Parametric equations are a set of equations in which x and y are defined as functions of another variable. The horizontal and vertical distance of a projectile are given as function of time, t, in seconds.

Apply

7. initial: $x = 0$, $y = 58$
At $t = 5$, $x = 181.2$, $y = 334.1$.

9.

t	x	y
0	0	58
2	72.5	264.5
4	144.9	342.9
6	217.4	293.4
8	289.9	115.8

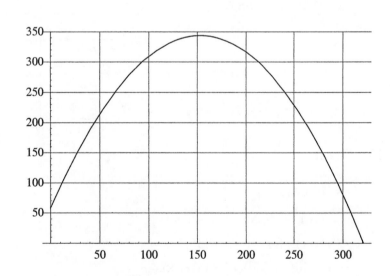

11. 1878.0 ft

$$x = 250(\cos 36°)t$$

$$y = -16t^2 + 250(\sin 36°)t + 15$$

When cannonball hits the ground, $y = 0$.

$$-16t^2 + 250(\sin 36°)t + 15 = 0$$

$$-16t^2 + 146.9463t + 15 = 0$$

$$t = \frac{-146.9463 \pm \sqrt{146.9463^2 - 4(-16)(15)}}{2(-16)}$$

$$t = \frac{-146.9463 \pm \sqrt{22553.2151}}{-32}$$

$$t = 9.2851$$

So, $x = 250(\cos 36°)t = 250(\cos 36°)(9.2851) = 1878.0 \text{ ft}$.

t	x	y
0	0	15
2	404.5	244.9
4	809.0	346.8
6	1213.5	320.7
8	1618.0	166.6

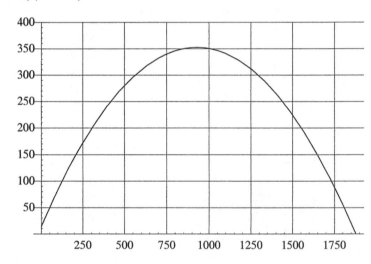

13. 55.4 ft

$$x = 40(\cos 39°)t$$

$$y = -16t^2 + 40(\sin 39°)t + 6 = 0$$

Solving for t, $t = 1.7836$.

$$x = 40(\cos 39°)(1.7836) = 55.4$$

15. 3.9 sec, 242.0 ft

$$60 \text{ mph} = 60\left(\frac{22}{15}\right) = 88 \text{ ft/sec}$$

$$x = 88(\cos 45°)t$$

$$y = -16t^2 + 88(\sin 45°)t = 0$$

Solving for t, $t = 3.8891$.

$$x = 88(\cos 45°)(3.8891) = 242.0$$

17. 3.8 sec, 317.3 ft

$$70 \text{ mph} = 80\left(\frac{22}{15}\right) = 102.6667 \text{ ft/sec}$$

$$x = 102.6667(\cos 36°)t$$

$$y = -16t^2 + 102.6667(\sin 36°)t + 3 = 0$$

Solving for t, $t = 3.8207$.

$$x = 102.6667(\cos 36°)(3.8207) = 317.3$$

Explore

19. Yes. The ball travels 523.5 ft in the horizontal direction.

$$90 \text{ mph} = 90\left(\frac{22}{15}\right) = 132 \text{ ft/sec}$$

$$x = 132(\cos 35°)t$$

$$y = -16t^2 + 132(\sin 35°)t + 8.5 = 0$$

Solving for t, $t = 4.8417$.

$$x = 132(\cos 35°)(4.8417) = 523.5$$

21. 2.2 sec, 32.0 mph

$$x = v(\cos 42°)t$$

$$75.8542 = 0.7431vt \quad \text{(note : } 75°10.25' = 75.8542°)$$

$$\frac{75.8542}{0.7431t} = v$$

$$y = -16t^2 + v(\sin 42°)t + 7.5$$

$$0 = -16t^2 + 0.6691\left(\frac{75.8542}{0.7431t}\right)t + 7.5$$

$$16t^2 = 75.8004$$

$$t = 2.1766$$

Thus, $v = \dfrac{75.8542}{0.7431t} = \dfrac{75.8542}{0.7431(2.1766)} = 46.9$ ft/sec.

$$46.9 \; \frac{\text{ft}}{\text{sec}} \div \frac{22}{15} = 32.0 \; \frac{\text{mi}}{\text{hr}}$$

23. 479.4 ft

$$50 \text{ mph} = 50\left(\frac{22}{15}\right) = 73.3333 \text{ ft/sec}$$

$$x = 73.3333(\cos 20°)t$$

$$y = -16t^2 + 73.3333(\sin 20°)t + 600 = 0$$

Solving for t, $t = 6.9575$.

$$x = 73.3333(\cos 20°)(6.9575) = 479.4$$

Review Section 8.1

1. 53 in.

$$c^2 = a^2 + b^2$$
$$c^2 = 28^2 + 45^2$$
$$c^2 = 2809$$
$$c = 53$$

2. 18.4 in.

$$x^2 + x^2 = 26^2$$
$$2x^2 = 676$$
$$x^2 = 338$$
$$x = 18.4$$

3. $\sin X = x/a$, $\cos X = t/a$, $\tan T = t/x$
   Use *soh cah toa*.

4. (a) 0.9205, (b) 0.9759

5. (a) 13.6°, (b) 88.4°

Review Section 8.2

6. $a = 35$, $\angle T = 36.9°$, $\angle X = 53.1°$

$$a^2 = t^2 + x^2$$
$$a^2 = 21^2 + 28^2$$
$$a^2 = 1225$$
$$a = 35$$

$$\tan T = \frac{21}{28}$$
$$\tan T = 0.75$$
$$T = \tan^{-1}(0.75)$$
$$T = 36.9°$$

$$\angle X = 90° - 36.9°$$
$$= 53.1°$$

7. $\angle U = 67°$, $u = 86.9$, $s = 94.4$

$$\angle U = 90° - 23°$$
$$= 67°$$

$$\tan 23° = \frac{36.9}{u}$$
$$0.4245 = \frac{36.9}{u}$$
$$0.4245u = 36.9$$
$$u = 86.9$$

$$\sin 23° = \frac{36.9}{s}$$
$$0.3907 = \frac{36.9}{s}$$
$$0.3907s = 36.9$$
$$s = 94.4$$

8. $\angle P = 16°$, $u = 98.5$, $p = 28.2$

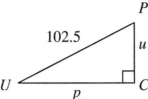

$\angle U = 74°$

$\angle P = 90° - 74°$
$\quad = 16°$

$\sin 74° = \dfrac{u}{102.5}$

$0.9613 = \dfrac{u}{102.5}$

$98.5 = u$

$\cos 74° = \dfrac{p}{102.5}$

$0.2756 = \dfrac{p}{102.5}$

$28.2 = p$

9. $90°$, $43.6°$, $46.4°$
   Since it is a right triangle, one angle is $90°$.

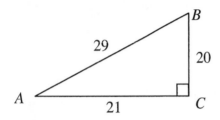

$\cos A = \dfrac{21}{29} = 0.7241$

$A = \cos^{-1}(0.7241) = 43.6°$

$B = 90° - 43.6° = 46.4°$

Review Section 8.3

10. The angle of elevation is measured upward from the horizontal while the angle of depression is measured downward from the horizontal.

11. 117.8 ft

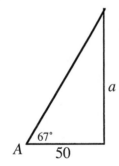

$\tan 67° = \dfrac{a}{50}$

$2.3559 = \dfrac{a}{50}$

$117.8 = a$

12. 4979.0 ft

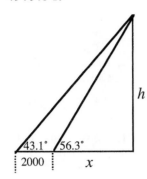

$$\tan 56.3° = \frac{h}{x}$$

$$1.4994 = \frac{h}{x}$$

$$1.4994x = h$$

$$x = \frac{h}{1.4994}$$

$$x = 0.6669h$$

$$\tan 43.1° = \frac{h}{x + 2000}$$

$$0.9358 = \frac{h}{x + 2000}$$

$$0.9358x + 1871.6 = h$$

Substituting $x = 0.6669h$

$$0.9358(0.6669h) + 1871.6 = h$$

$$0.6241h + 1871.6 = h$$

$$1871.6 = 0.3759h$$

$$4979.0 = h.$$

Review Section 8.4

13. $\angle A = 83°$, $a = 7.2$, $s = 4.2$

$$\angle A = 180° - 35° - 62° = 83°$$

$$\frac{6.4}{\sin 62°} = \frac{a}{\sin 83°}$$

$$\frac{6.4}{0.8829} = \frac{a}{0.9925}$$

$$0.8829a = 6.352$$

$$a = 7.2$$

$$\frac{6.4}{\sin 62°} = \frac{s}{\sin 35°}$$

$$\frac{6.4}{0.8829} = \frac{s}{0.5736}$$

$$0.8829s = 3.6710$$

$$s = 4.2$$

14. $\angle M = 46°$, $\angle T = 46°$, $a = 19.5$

$$a^2 = 14^2 + 14^2 - 2(14)(14)\cos 88°$$

$$a^2 = 196 + 196 - 392(0.0349)$$

$$a^2 = 378.3192$$

$$a = 19.5$$

Since $t = m$, $\angle T = \angle M$ (an isosceles triangle)

$$\angle M + \angle M + 88° = 180°$$

$$2\angle M = 92°$$

$$\angle M = 46°$$

and $\angle T = 46°$

15. $\angle P = 80.7°,\ c = 44.4,\ a = 99.8$

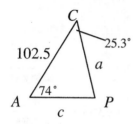

$$\angle P = 180° - 74° - 25.3° = 80.7°$$

$$\frac{102.5}{\sin 80.7°} = \frac{c}{\sin 25.3°}$$

$$\frac{102.5}{0.9869} = \frac{c}{0.4274}$$

$$0.9869c = 43.8085$$

$$c = 44.4$$

$$\frac{102.5}{\sin 80.7°} = \frac{a}{\sin 74°}$$

$$\frac{102.5}{0.9869} = \frac{a}{0.9613}$$

$$0.9869a = 98.5333$$

$$a = 99.8$$

16. $\angle N = 65.0°,\ \angle A = 31.2°,\ \angle P = 83.8°$

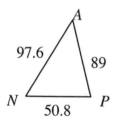

$$89^2 = 97.6^2 + 50.8^2 - 2(97.6)(50.8)\cos N$$

$$7921 = 9525.76 + 2580.64 - 9916.16\cos N$$

$$-4185.4 = -9916.16\cos N$$

$$0.4221 = \cos N$$

$$65.0° = N$$

$$50.8^2 = 97.6^2 + 89^2 - 2(97.6)(89)\cos N$$

$$2580.64 = 9525.76 + 7921 - 17{,}372.8\cos A$$

$$-14{,}866.12 = -17{,}372.8\cos A$$

$$0.8557 = \cos A$$

$$31.2° = A$$

$$\angle P = 180° - 65.0° - 31.2° = 83.8°$$

17. 3.7 mi

   To solve this problem, determine the distance from the bear to Station B on both
   Monday and Tuesday. The difference between these distances is the solution. After
   determining the angle at C on each day, we use the Law of Sines.

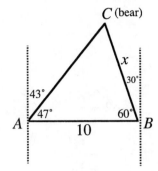

On Monday, $\angle C = 180° - 47° - 60° = 73°$

$$\frac{x}{\sin 47°} = \frac{10}{\sin 73°}$$

$$\frac{x}{0.7314} = \frac{10}{0.9563}$$

$$x = 7.6$$

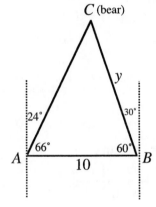

On Tuesday, $\angle C = 180° - 66° - 60° = 54°$

$$\frac{y}{\sin 66°} = \frac{10}{\sin 54°}$$

$$\frac{y}{0.9135} = \frac{10}{0.8090}$$

$$y = 11.3$$

$$y - x = 11.3 - 7.6 = 3.7 \text{ mi}$$

18. 213.0 ft

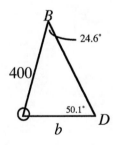

$$\frac{b}{\sin 24.6°} = \frac{400}{\sin 50.1°}$$

$$\frac{b}{0.4163} = \frac{400}{0.7672}$$

$$b = 217.0$$

Actual distance $= 217.0 - 4 = 213.0$

19. 22.2 mi, 13.7 mi, 12.1 mi

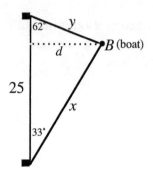

$$B = 180° - 62° - 33° = 85°$$

$$\frac{x}{\sin 62°} = \frac{25}{\sin 85°}$$

$$\frac{x}{0.8829} = \frac{25}{0.9962}$$

$$x = 22.2$$

$$\frac{y}{\sin 33°} = \frac{25}{\sin 85°}$$

$$\frac{y}{0.5446} = \frac{25}{0.9962}$$

$$y = 13.7$$

$$\sin 62° = \frac{d}{y}$$

$$0.8829 = \frac{d}{13.7}$$

$$12.1 = d$$

20. 2950.8 yd, 2987.6 yd

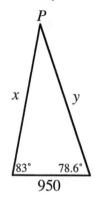

$$P = 180° - 83° - 78.6° = 18.4°$$

$$\frac{x}{\sin 78.6°} = \frac{950}{\sin 18.4°}$$

$$\frac{x}{0.9803} = \frac{950}{0.3156}$$

$$x = 2950.8$$

$$\frac{y}{\sin 83°} = \frac{950}{\sin 18.4°}$$

$$\frac{y}{0.9925} = \frac{950}{0.3156}$$

$$y = 2987.6$$

Review Section 8.6

21. At $t = 0$, $x = 0$, $y = 10$

At $t = 4$, $x = 201.9$, $y = 42.3$

$$x = 88(\cos 55°)t$$

$$y = -16t^2 + 88(\sin 55°)t + 10$$

At $t = 0$, $\begin{cases} x = 88(\cos 55°)(0) = 0 \\ y = -16(0)^2 + 88(\sin 55°)(0) + 10 = 10 \end{cases}$

At $t = 4$, $\begin{cases} x = 88(\cos 55°)(4) = 201.9 \\ y = -16(4)^2 + 88(\sin 55°)(4) + 10 = 42.3 \end{cases}$

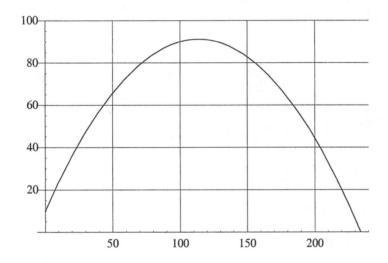

22. 67.6 ft

$$50 \text{ mph} = 50\left(\frac{22}{15}\right) \text{ft/sec} = 73.3333 \text{ ft/sec}$$

$$x = 73.3333(\cos 32°)t$$

$$y = -16t^2 + 73.3333(\sin 32°)t + 20 = 0$$

$$-16t^2 + 38.8607t + 20 = 0$$

$$t = \frac{-38.8607 \pm \sqrt{38.8607^2 - 4(-16)20}}{2(-16)}$$

$$t = 2.8651$$

$$x = 73.3333(\cos 32°)(2.8651) = 178.2$$

$$60 \text{ mph} = 60\left(\frac{22}{15}\right)\text{ft/sec} = 88 \text{ ft/sec}$$

$$x = 88(\cos 32°)t$$
$$y = -16t^2 + 88(\sin 32°)t + 20 = 0$$
$$-16t^2 + 46.6329t + 20 = 0$$
$$t = \frac{-46.6329 \pm \sqrt{46.6329^2 - 4(-16)20}}{2(-16)}$$
$$t = 3.2940$$
$$x = 88(\cos 32°)(3.2940) = 245.8$$

Difference: $245.8 - 178.2 = 67.6$ ft.

23. 44.0 ft/sec

$$x = v(\cos 42°)t$$
$$x = 0.7431vt$$
$$66.2 = 0.7431vt$$
$$\frac{66.2}{0.7431t} = v$$
$$y = -16t^2 + v(\sin 42°)t + 6$$
$$0 = -16t^2 + \left(\frac{66.2}{0.7431t}\right)(0.6691)t + 6$$
$$16t^2 = 65.6040$$
$$t^2 = 4.1003$$
$$t = 2.0249$$

Thus, $v = \dfrac{89.0809}{t} = \dfrac{89.0809}{2.0249} = 44.0$ ft/sec

Chapter 8 Test

1. $b = 51$, $\angle A = 28.1°$, $\angle C = 61.9°$

$$b^2 = 24^2 + 45^2$$
$$= 576 + 2025$$
$$= 2601$$
$$b = 51$$

$$\sin A = \frac{24}{51}$$
$$\sin A = 0.4706$$
$$A = 28.1°$$
$$C = 90° - 28.1° = 61.9°$$

2. $\angle N = 84°$, $r = 20.3$, $n = 29.0$

$$N = 180° - 44° - 52° = 84°$$

$$\frac{r}{\sin 44°} = \frac{23}{\sin 52°}$$

$$\frac{r}{0.6947} = \frac{23}{0.7880}$$

$$0.7880r = 15.9781$$

$$r = 20.3$$

$$\frac{n}{\sin 84°} = \frac{23}{\sin 52°}$$

$$\frac{n}{0.9945} = \frac{23}{0.7880}$$

$$0.7880n = 22.8735$$

$$n = 29.0$$

3. $a = 38.0$, $\angle M = 73.1°$, $\angle Y = 44.4°$

$$a^2 = 30^2 + 41^2 - 2(30)(41)\cos 62.5°$$

$$a^2 = 900 + 1681 - 1135.9016$$

$$a^2 = 1445.0984$$

$$a = 38.0$$

$$41^2 = 30^2 + 38^2 - 2(30)(38)\cos M$$

$$1681 = 900 + 1444 - 2280\cos M$$

$$0.2908 = \cos M$$

$$73.1° = M$$

$$Y = 180° - 62.5° - 73.1° = 44.4°$$

4. $a = 45.7$

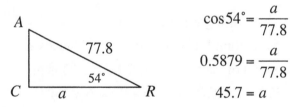

$$\cos 54° = \frac{a}{77.8}$$

$$0.5879 = \frac{a}{77.8}$$

$$45.7 = a$$

5. $\angle C = 84.9°$

$$110^2 = 73^2 + 89^2 - 2(73)(89)\cos C$$

$$12{,}100 = 5329 + 7921 - 12{,}994\cos C$$

$$-1150 = -12{,}994\cos C$$

$$0.0885 = \cos C$$

$$84.9° = C$$

6. $41.8°$

$$\sin A = \frac{20}{30}$$

$$\sin A = 0.6667$$

$$A = 41.8°$$

7. (a) 17.0 ft, 176.2 ft

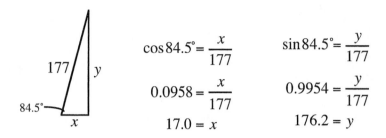

$$\cos 84.5° = \frac{x}{177} \qquad \sin 84.5° = \frac{y}{177}$$

$$0.0958 = \frac{x}{177} \qquad 0.9954 = \frac{y}{177}$$

$$17.0 = x \qquad 176.2 = y$$

(b) 48.0 ft, 178.9 ft

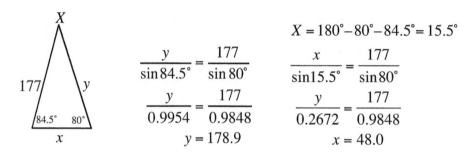

$$X = 180° - 80° - 84.5° = 15.5°$$

$$\frac{y}{\sin 84.5°} = \frac{177}{\sin 80°} \qquad \frac{x}{\sin 15.5°} = \frac{177}{\sin 80°}$$

$$\frac{y}{0.9954} = \frac{177}{0.9848} \qquad \frac{y}{0.2672} = \frac{177}{0.9848}$$

$$y = 178.9 \qquad x = 48.0$$

8. 851.8 mi

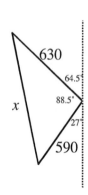

The angle between the two flight paths is $180° - 62.5° - 27° = 88.5°$. Since the planes are traveling for two hours, the distances are found by multiplying the speed of each plane by 2 hours. This gives $d_1 = (295)(2) = 590$ mi and $d_2 = (315)(2) = 630$ mi. Using the Law of Cosines gives the following.

$$x^2 = 590^2 + 630^2 - 2(590)(630)\cos 88.5°$$

$$x^2 = 348,100 + 396,900 - 19,459.9434$$

$$x = 851.8$$

9. N11.2°E, N48.2°W

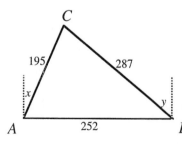

In the diagram, the bearing from Island A to Island C is labeled x and the bearing from Island B to Island C is labeled y. To determine the bearings, we first find $\angle A$ and $\angle B$ by using the Law of Cosines.

$$287^2 = 252^2 + 195^2 - 2(252)(195)\cos A$$

$$82,369 = 63,504 + 38,025 - 98,280\cos A$$

$$0.1950 = \cos A$$

$$78.8° = A$$

Bearing $x = 90° - 78.8° = 11.2° = \text{N}11.2°\text{E}$

$$195^2 = 252^2 + 287^2 - 2(252)(287)\cos B$$

$$38,025 = 63,504 + 82,369 - 144,648\cos B$$

$$0.7456 = \cos B$$

$$41.8° = B$$

Bearing $y = 90° - 41.8° = 48.2° = \text{N}48.2°\text{W}$

10. 319.0 ft

$$60 \text{ mph} = 60\left(\frac{22}{15}\right) = 88 \text{ ft/sec}$$

$$x = 88(\cos 35°)t$$

$$y = -16t^2 + 88(\sin 35°)t + 90 = 0$$

Solving for t, $t = 4.4257$.

$$x = 88(\cos 35°)(4.4257) = 319.0$$

11. 46.9 ft/sec

$$x = v(\cos 40°)t$$

$$x = 0.7660vt$$

$$74.25 = 0.7660vt$$

$$\frac{74.25}{0.7660t} = v$$

$$y = -16t^2 + v(\sin 40°)t + 6$$

$$= -16t^2 + 0.6428vt + 6$$

$$0 = -16t^2 + 0.6428\left(\frac{74.25}{0.7660t}\right)t + 6$$

$$16t^2 = 68.3080$$

$$t^2 = 4.2693$$

$$t = 2.0662$$

Thus, $v = \dfrac{74.25}{0.7660t} = \dfrac{74.25}{0.7660(2.0662)} = 46.9.$

CHAPTER 9   MATH OF FINANCE

Handling financial dealings is one of the responsibilities of an adult in today's society. In this chapter, you will learn how to calculate percentage increase and decrease and determine the percentage markup or markdown on a sale. You will determine the amount of interest earned in various types of accounts, using both simple and compound interest. You will learn how to calculate the value of an annuity and the amount of a loan payment and to see the benefits of paying off a loan early. If the you ever plan to buy a car or save for retirement, this chapter will pique your interest.

SECTION 9.1   PERCENT

This section discusses working with percents, including markups, markdowns, and percentage increases and decreases. The single formula presented in this section is used to calculate percentage increase or decrease.

$$\textbf{Percent increase or decrease } = \frac{A}{B} \text{ where } \begin{cases} A = \text{amount of increase or decrease} \\ B = \text{base (starting) amount} \end{cases}$$

Note that the decimal for $\dfrac{A}{B}$ is converted to a percent.

Explain

1. Percent means out of one hundred.

3. To change a percent to a decimal, drop the percent sign and divide by 100.

5. Multiply the price of the item by 0.052 and then add this amount to the price of the item. You could also simply multiply the price of the item by 1.052.

7. A markup is an increase in the price of an item. A markdown is a decrease in the price of an item. Both are calculated by multiplying the percentage markup or markdown by the original price of the item and then adding/subtracting it from the original price.

9. In determining the percent decrease in enrollment, say for a year, you need to know the enrollment at the beginning of the year and the enrollment at the end of the year. To calculate the percent increase, subtract the end of the year enrollment from the beginning of the year enrollment, divide this amount by the beginning of the year enrollment and convert the decimal to a percent.

Apply

11. (a) $5 \div 100 = 0.05$ (b) $6.7 \div 100 = 0.067$ (c) $9.125 \div 100 = 0.09125$
    (d) $234 \div 100 = 2.34$ (e) $0.03 \div 100 = 0.0003$

13. (a) $0.35 \times 100 = 35\%$ (b) $0.06 \times 100 = 6\%$ (c) $0.0025 \times 100 = 0.25\%$
    (d) $2.015 \times 100 = 201.5\%$ (e) $0.75 \times 100 = 75\%$

15. (a) $5 \div 7 \approx 0.714 = 71.4\%$ (b) $2 \div 3 = 0.667 = 66.7\%$ (c) $16 \div 29 \approx 0.552 = 55.2\%$
    (d) $34 \div 995 \approx 0.034 = 3.4\%$ (e) $5 \div 64 \approx 0.078 = 7.8\%$

17. (a) $34 \times 0.07 = \$2.38$ (b) $125.95 \times 0.07 = \$8.82$ (c) $675.79 \times 0.07 = \$47.31$

19. (a) $17.50 = P - 0.15P$ (b) $17.50 = P - 0.40P$ (c) $17.50 = P - 0.75P$

$17.50 = 0.85P$ $17.50 = 0.60P$ $17.50 = 0.25P$

$$P = \frac{17.50}{0.85} = \$20.59 \qquad P = \frac{17.50}{0.60} = \$29.17 \qquad P = \frac{17.50}{0.25} = \$70.00$$

21. (a) $A = 20 + 0.35 \times 20 = \$27.00$
    (b) $A = 45.60 + 0.35 \times 45.60 = \$61.56$
    (c) $A = 109.99 + 0.35 \times 109.99 = \$148.49$

23. (a) $17.00 = P + 0.25P$ (b) $25.50 = P + 0.25P$ (c) $29.99 = P + 0.25P$

$17.00 = 1.25P$ $25.50 = 1.25P$ $29.99 = 1.25P$

$$P = \frac{17.00}{1.25} = \$13.60 \qquad P = \frac{25.50}{1.25} = \$20.40 \qquad P = \frac{29.99}{1.25} = \$23.99$$

25. (a) $\dfrac{9.00 - 7.79}{9.00} = 0.134 = 13.4\%$ (b) $\dfrac{432 - 216.50}{432} = 0.499 = 49.9\%$

    (c) $\dfrac{4999 - 3000}{4999} = 0.400 = 40.0\%$

Explore

27. $379.99 = P - 0.003P$

$379.99 = 0.997P$

$$P = \frac{379.99}{0.997} = \$381.13$$

29. (a) $A = 55 + 0.40 \times 55 = \$77.00$
    (b) sale price $= 77 - 0.40 \times 77 = \$46.20$. Therefore, the outlet loses $55 - 46.20 = \$8.80$.

31. $6780 + 0.275(105,000 - 45,200) = 6780 + 0.275 \times 59,800 = 6780 + 16,445 = \$23,225$

33. (a) Mitek lost $\dfrac{37\frac{3}{8} - 29\frac{1}{2}}{37\frac{3}{8}} = \dfrac{37.375 - 29.5}{37.375} = 0.211 = 21.1\%$

    (b) Yotek lost $\dfrac{8\frac{5}{8} - 6\frac{1}{4}}{8\frac{5}{8}} = \dfrac{8.625 - 6.25}{8.625} = 0.275 = 27.5\%$

Therefore, Yotek had the greater percentage decrease.

35. (a) Discount $= 0.085 \times 4200 = \$357$

(b) Discount $= 0.085 \times 5000 + 0.095 \times (9890 - 5000) = 425 + 464.55 = \$889.55$

(c) Discount $= 0.085 \times 5000 + 0.095 \times (10{,}000) + 0.105 \times (19{,}750 - 15{,}000)$
$$= 425 + 950 + 498.75 = \$1873.75$$

SECTION 9.2 SIMPLE INTEREST

This section discusses determining simple interest. Simple interest is the method of interest calculation where the interest is paid at the end of the specified time and is earned on the principal only. The two formulas in the section are:

$$I = Prt \quad \text{and} \quad A = P(1 + rt) \quad \text{where} \quad \begin{cases} I = \text{interest} \\ P = \text{principal} \\ r = \text{interest rate} \\ t = \text{time} \\ A = \text{amount in the account} \end{cases}$$

It is important to make sure that the time (t) is measured in the same unit of time as the interest rate (r). For example, suppose the interest rate is 12% per year and the time is 18 months. Before using the formulas, the time can be converted from 18 months to 1.5 years, or the rate can be converted to 1% per month.

Explain

1. Principal is the amount deposited into an account.

3. The length of time in years of an investment can be determined by multiplying the principal by the annual interest rate and dividing this result into the amount of interest.

5. In order to determine the interest, the interest rate and the length of time must be in the same units of time. Therefore if the rate is given as a monthly rate and the time is given in years, one possible method of determining the interest is to multiply the monthly interest rate by 12, multiply this rate by the number of years and then multiply this result by the principal.

Apply

7. $I = Prt = 2000(0.05)(4) = \$400.00$

9. $I = 2500(0.0075)(3) = \$56.25$

11. $A = P(1 + rt)$
$A = 3000[1 + (0.04)(6)]$
$A = \$3720.00$

13. Using $r = 0.04/12 = 0.003333$,
$$A = 35{,}000\left[1 + (0.003333)(4)\right]$$
$$A = \$35{,}466.67$$

15. Using $t = 1 \times 365 = 365$,
$$A = 1000\left[1 + (0.000247)(365)\right]$$
$$A = \$1090.16$$

17. $3100 = P\left[1 + (0.04)(6)\right]$
$$3100 = 1.24P$$
$$P = \frac{3100}{1.24}$$
$$P = \$2500.00$$

19. $3300 = 3000[1 + 0.04t]$
$$1.1 = 1 + 0.04t$$
$$0.1 = 0.04t$$
$$t = \frac{0.1}{0.04} = 2.5 \text{ years}$$

21. $3000 = 2000[1 + 2r]$
$$1.5 = 1 + 2r$$
$$0.5 = 2r$$
$$r = \frac{0.5}{2} = 0.25 \text{ or } 25\% \text{ per year}$$

Explore

23. $A = 3400\left[1 + (0.063)\left(\frac{8}{12}\right)\right]$
$$A = \$3542.80$$

25. $\quad 5000 = 4700[1 + 0.0551t]$
$$1.063830 = 1 + 0.0551t$$
$$0.063830 = 0.0551t$$
$$t = \frac{0.063830}{0.0551} = 1.16 \text{ years}$$

27. Purchase price = 2000 × 87.88 = $175,760
     Sale price = 2000 × 93.12 = $186,240
     Using $t = 18/12 = 1.5$,
     $$186{,}240 = 175{,}760[1 + 1.5r]$$
     $$1.059627 = 1 + 1.5r$$
     $$0.059627 = 1.5r$$
     $$r = \frac{0.059627}{1.5} = 0.0398 \text{ or } 3.98\% \text{ per year.}$$

29. $$20{,}000 = 18{,}500[1 + r]$$
     $$1.081081 = 1 + r$$
     $$r = 0.081081 \text{ or } 8.11\% \text{ per year}$$

31. $$50{,}000 = 31{,}250[1 + 10r]$$
     $$1.6 = 1 + 10r$$
     $$0.6 = 10r$$
     $$r = 0.06 \text{ or } 6.0\% \text{ per year}$$

33. $$1{,}000{,}000 = 200{,}000(1 + 0.085t)$$
     $$1 + 0.085t = \frac{1{,}000{,}000}{200{,}000} = 5$$
     $$0.085t = 4$$
     $$t = \frac{4}{0.085} = 47.1 \text{ years}$$

35. (a) $1000(1.025) = \$1025.00$
     (b) $1025(1.025) = \$1050.63$
     (c) $1050.63(1.025) = \$1076.90$

36. $5.82

37. $920 \times .00003833 \times 30 = \$10.58$

Compound interest is a method of interest calculation where the interest is paid at regular intervals on both the principal and any previously accumulated interest.

The **compound interest formula** is:

$$A = P(1 + r)^n \quad \text{where} \begin{cases} A = \text{amount in the account after } n \text{ time periods} \\ P = \text{the amount deposited} \\ r = \text{annual interest rate} \div \text{number of periods per year} \\ n = \text{number of years} \times \text{number of periods per year} \end{cases}$$

An important note:
Since time is represented in the equation by the exponent n, if the problem involves solving for time, you should use logarithms.

Explain

1. To find the amount in an account earning compound interest, you need to know the principal (P), the interest rate (r), and the number of periods (n).

3. The compound interest method accumulates more interest because, during each year, interest is earned on the principal and on the interest earned in the previous years. Using simple interest, interest is earned on only the principal at the end of a given time period.

5. When compound interest is computed, interest is added at the end of every period. To find the number of periods, multiply the number of years by the number of periods per year.

7. The effective interest rate is the simple interest rate which, over a one year period, provides the same amount of interest as the compound rate.

Apply

9. Using $r = 0.06/12 = 0.005$, $n = 4(12) = 48$,

$$A = P(1 + r)^n$$

$$A = 2000(1.005)^{48}$$

$$A = \$2540.98$$

11. $r = 0.05/365 = 0.000136986$, $n = 3(365) = 1095$,

$$A = 6000(1.000136986)^{1095}$$

$$A = \$6970.93$$

13. $r = 0.06/12 = 0.005$, $n = 2(12) = 24$,

$$6000 = P(1 + 0.005)^{24}$$

$$6000 = P(1.127160)$$

$$P = \$5323.11$$

15. $r = 0.06/4 = 0.015$, $n = 5(4) = 20$,

$$7500 = P(1 + 0.015)^{20}$$
$$7500 = P(1.346855)$$
$$P = \$5568.53$$

17. $r = 0.09/12 = 0.0075$, $n = 12t$,

$$5000 = 3500(1 + 0.0075)^{12t}$$
$$1.428571 = 1.0075^{12t}$$
$$\ln(1.428571) = 12t \ln(1.0075)$$
$$t = \frac{\ln(1.428571)}{12 \ln(1.0075)} \approx 3.98 \text{ years}$$

19. $r = 0.078/12 = 0.0065$, $n = 12t$,

$$15,000 = 5000(1 + 0.0065)^{12t}$$
$$3 = 1.0195^{12t}$$
$$\ln(3) = 12t \ln(1.0065)$$
$$t = \frac{\ln(3)}{12 \ln(1.0065)} \approx 14.13 \text{ years}$$

21.

$$5000 = 3500(1 + r)^2$$
$$1.428571 \approx (1 + r)^2$$
$$1.428571^{1/2} \approx \left[(1 + r)^2\right]^{1/2}$$
$$1.195228 \approx 1 + r$$
$$r \approx 0.195228 \approx 19.52\%$$

23. The rate is compounded monthly so $n = 12(15) = 180$.

$$75,000 = 25,000(1 + r)^{180}$$
$$3 \approx (1 + r)^{180}$$
$$3^{1/180} \approx \left[(1 + r)^{180}\right]^{1/180}$$
$$1.006122 \approx 1 + r$$
$$r \approx 0.006122$$

Since the rate is compounded monthly, the annual rate is $12(0.006122) \approx 7.35\%$.

25. $r_{eff} = \left(1 + \dfrac{0.06}{12}\right)^{12} - 1 \approx 0.0617 = 6.17\%$

27. $r_{eff} = \left(1 + \dfrac{0.13}{52}\right)^{52} - 1 \approx 0.1386 = 13.86\%$

Explore

29. At 8% compounded annually, $A = 1000(1.08) = \$1080$.
    At 7.8% compounded daily, $r = 0.078/365 = 0.000213699$ and $n = 365$. This gives $A = 1000(1.000213699)^{365} = \$1081.11$. Thus, the 7.8%, compounded daily is a better investment.

31. $r = 0.066/4 = 0.0165$, $n = 4(5) = 20$,
    $A = 15{,}000(1.0165)^{20} = \$20{,}808.41$

33. (a) $r = 0.05475/365 = 0.00015$, $n = 365$,
    $A = 10{,}000(1.00015)^{365} = \$10{,}562.72$
    Since the principal is \$10,000, the interest in the first year is
    $10{,}562.72 - 10{,}000 = \$562.72$.

    (b) $n = 2(365) = 730$,
    $A = 10{,}000(1.00015)^{730} = \$11{,}157.11$.
    Since the principal at the beginning of the second year is \$10,562.72, the interest earned in the second year is $11{,}157.11 - 10{,}562.72 = \$594.39$.

    (c) $n = 3(365) = 1095$,
    $A = 10{,}000(1.00015)^{1095} = \$11{,}784.94$.
    Since the principal at the beginning of the third year is \$11,157.11, the interest earned in the third year is $11{,}784.94 - 11{,}157.11 = \$627.83$.

35. Using $n = 12(5) = 60$ and $r = 0.09/12 = 0.0075$
    $15{,}000 = P(1 + 0.0075)^{60}$

    $P = \dfrac{15{,}000}{1.0075^{60}} = \$9580.50$.

37. The account starts with $r = 0.09/12 = 0.0075$ and $n = 12(8) = 96$. Thus, after the first 8 years, the account is worth $A = 10{,}000(1.0075)^{96} = \$20{,}489.21$.

    Next, a new account is created with $P = \$20{,}489.21$ and $r = 0.10/4 = 0.025$. We want $A = 1{,}000{,}000$. Solving for n, we have
    $1{,}000{,}000 = 20{,}489.21(1.025)^{n}$

    $1.025^{n} = \dfrac{1{,}000{,}000}{20{,}489.21} \approx 48.806177$

    $n = \dfrac{\ln(48.806177)}{\ln(1.025)} \approx 157.45$ quarters.

    Therefore, it will take approximately $157.45 \div 12 = 13$ years and two months.

39. Using $A = 40{,}000$, $P = 20{,}000$, $r = 0.12/365 = 0.000328767$, we have

$$40{,}000 = 20{,}000(1.000328767)^n$$

$$2 = 1.000328767^n$$

$$\ln(2) = n\ln(1.000328767)$$

$$n = \frac{\ln(2)}{\ln(1.000328767)} \approx 2109 \text{ days}$$

$$n \approx \frac{2109}{365} \approx 5.78 \text{ years}.$$

41. $P = 24$, $r = 0.05/365 = 0.000136986$, $n = 365(400) = 146{,}000$

$$A = 24(1.000136986)^{146{,}000} \approx \$11{,}628{,}024{,}100$$

43. In the first thirteen day period, $P = 3510.24$, $r = 0.0002164$, and $n = 13$. The balance is $A = 3510.24(1 + 0.0002164)^{13} = \$3520.13$ and the interest is $I = 3520.13 - 3510.24 = \$9.89$.

In the following six day period, $P = 3520.13 - 700 = \$2820.13$ and $n = 6$. The balance is $A = 2820.13(1 + 0.0002164)^{6} = \$2823.79$ and the interest is $I = 2823.79 - 2820.13 = \$3.66$.

In the final eleven day period, $P = 2823.79 + 120.50 = \$2944.29$ and $n = 11$. The balance is $A = 2944.29(1 + 0.0002164)^{11} = \$2951.31$ and the interest is $I = 2951.31 - 2944.29 = \$7.02$.

Thus, the total finance charge for the period is $\$9.89 + \$3.66 + \$7.02 = \$20.57$.

45. In the first eleven day period $P = 2153.64$, $r = 0.0002164$, and $n = 11$. The balance is

$$A = 2153.64(1 + 0.0002164)^{11} = \$2158.77$$

and the interest is $I = 2158.77 - 2153.64 = \$5.13$.

In the following eight day period, $P = 2158.77 - 900 = \$1258.77$ and $n = 8$. The balance is $A = 1258.77(1 + 0.0002164)^{8} = \$1260.95$ and the interest is $I = 1260.95 - 1258.77 = \$2.18$.

In the final eleven day period, $P = 1260.95 + 120.50 = \$1381.10$ and $n = 11$. The

balance is $A = 1381.10(1 + 0.0002164)^{11} = \$1384.39$

and the interest is $I = 1384.39 - 1381.10 = \$3.29$.

Thus, the total finance charge for the period is $\$5.13 + \$2.18 + \$3.29 = \$10.60$.

SECTION 9.4 ANNUITIES

An annuity is an account into which money is deposited on a regular basis.

The annuity formula is:

$$S = PMT\left[\frac{(1+r)^n - 1}{r}\right] \text{ where } \begin{cases} S = \text{amount in the annuity} \\ PMT = \text{the amount of each payment (deposit)} \\ r = \text{annual interest rate} \div \text{number of periods per year} \\ n = \text{number of years} \times \text{number of periods per year} \end{cases}$$

The difference between an annuity and a compound interest account is that with an annuity, regular deposits are made for the life of the annuity as opposed to the single deposit made into a compound interest account.

Explain

1. An ordinary annuity is an account at a fixed rate of interest into which a fixed amount of money is deposited at the end of each period.

3. To determine the total amount of interest earned by an annuity, determine the total value of the annuity and subtract the total of the payments made into the annuity.

Apply

5. Using $r = 0.06/12 = 0.005$ and $n = 12(5) = 60$,

$$S = PMT\left[\frac{(1+r)^n - 1}{r}\right]$$

$$S = 200\left[\frac{(1+0.005)^{60} - 1}{0.005}\right] = \$13,954.01.$$

7. Using $r = 0.06/4 = 0.015$ and $n = 4(25) = 100$,

$$S = 150\left[\frac{(1+0.015)^{100} - 1}{0.015}\right] = \$34,320.46.$$

9. Using $r = 0.04/4 = 0.01$ and $n = 4(10) = 40$,

$$20{,}000 = PMT\left[\frac{(1+0.01)^{40} - 1}{0.01}\right]$$

$$20{,}000 = PMT(48.886373)$$

$$PMT = \$409.11 \text{ per month.}$$

11. Using $r = 0.06/12 = 0.005$ and $n = 12(20)$,

$$200{,}000 = PMT\left[\frac{(1+0.005)^{240} - 1}{0.005}\right]$$

$$200{,}000 = PMT(462.040895)$$

$$PMT = \$432.86.$$

13. Using $r = 0.09/12 = 0.075$,

$$20{,}000 = 100\left[\frac{(1+0.0075)^{n} - 1}{0.0075}\right]$$

$$200 = \left[\frac{(1+0.0075)^{n} - 1}{0.0075}\right]$$

$$1.5 = 1.0075^{n} - 1$$

$$2.5 = 1.0075^{n}$$

$$\ln 2.5 = n \ln 1.0075$$

$$n = \frac{\ln 2.5}{\ln 1.0075} \approx 123.$$

Therefore, $t \approx 123$ months or 10.22 years.

15. Using $r = 0.072/12 = 0.006$,

$$20{,}000 = 50\left[\frac{(1+0.006)^{n} - 1}{0.006}\right]$$

$$400 = \left[\frac{(1+0.006)^{n} - 1}{0.006}\right]$$

$$2.4 = 1.006^{n} - 1$$

$$3.4 = 1.006^{n}$$

$$\ln(3.4) = n \ln(1.06)$$

$$n = \frac{\ln(3.4)}{\ln(1.006)} \approx 205.$$

Therefore, $t \approx 205$ months or 17.08 years.

Explore

17. Using $r = 0.093/4 = 0.02325$ and $n = 4(25) = 100$,

$$S = 500\left[\frac{(1 + 0.02325)^{100} - 1}{0.02325}\right] = \$192{,}646.74.$$

19. (a) Using $r = 0.09/12 = 0.0075$ and $n = 12(48) = 576$,

$$S = 100\left[\frac{(1 + 0.0075)^{576} - 1}{0.0075}\right] = \$973{,}151.26.$$

(b) $576 \times 100 = \$57{,}600.00$

(c) $973{,}151.26 - 57{,}600 = \$915{,}551.26$

(d) Using $r = 0.09/12 = 0.0075$ and $n = 12(38) = 456$,

$$S = 200\left[\frac{(1 + 0.0075)^{456} - 1}{0.0075}\right] = \$778{,}181.07.$$

(e) $456 \times 200 = \$91{,}200.00$

(f) $778{,}181.07 - 91{,}200 = \$686{,}981.07$

21. (a) Using $r = 0.09/12 = 0.0075$ and $n = 12(10) = 120$,

$$S = 400\left[\frac{(1 + 0.0075)^{120} - 1}{0.0075}\right] = \$77{,}405.71.$$

(b) Using $r = 0.11/4 = 0.0275$ and $n = 4(25) = 100$,

$A = 77{,}405.71(1 + 0.0275)^{100} = \$1{,}166{,}691.60.$

23. Using $r = 0.06/12 = 0.005$, $PMT = 150$, and $S = 58{,}000$,

$$58{,}000 = 150\left[\frac{(1 + 0.005)^{n} - 1}{0.005}\right]$$

$$\frac{1.005^{n} - 1}{0.005} = \frac{58{,}000}{150} = 386.666667$$

$$1.005^{n} = 2.933333$$

$$n = \frac{\ln(2.933333)}{\ln(1.005)} \approx 215.77$$

$$t \approx 18 \text{ years.}$$

25. Using $r = 0.063/12 = 0.00525$ and $n = 12(20) = 240$,

$$1{,}000{,}000 = PMT\left[\frac{(1 + 0.00525)^{240} - 1}{0.00525}\right]$$

$$1{,}000{,}000 = PMT(478.823035)$$

$$PMT = \$2088.45 \text{ per month.}$$

27. (a) Using $r = 0.078/12 = 0.0065$ and $n = 12(30) = 360$,

$$200,000 = PMT\left[\frac{(1+0.0065)^{360} - 1}{0.0065}\right]$$

$$200,000 = PMT(1431.219167)$$

$$PMT = \$139.74 \text{ per month.}$$

(b) After 15 years, $n = 12(15) = 180$,

$$S = 139.74\left[\frac{(1+0.0065)^{180} - 1}{0.0065}\right] = \$47,507.66.$$

SECTION 9.5 LOANS

The loans discussed in this section have equal payments each period. Each payment will include both principal and interest.

The loan formula is:

$$L = PMT\left[\frac{1 - (1+r)^{-n}}{r}\right] \quad \text{where} \quad \begin{cases} L = \text{amount of the loan} \\ PMT = \text{the amount of each payment} \\ r = \text{annual interest rate} \div \text{ number of periods per year} \\ n = \text{number of years} \times \text{ number of periods per year} \end{cases}$$

Explain

1. The amount of interest in each payment does not remain the same; it decreases. The reason that the interest included in each payment decreases is because the interest is based on the balance of the loan and the balance of the loan decreases with each payment.

3. Accelerating the payments on a loan means that you are paying more than is required under the terms of the loan. The result of accelerating the payments is that the loan is paid off in a shorter length of time and that the total amount of interest paid on the loan is less than the amount of interest paid if the loan is not accelerated.

5. When comparing renting to buying, you should look at the monthly rental cost versus the mortgage payments, points, and tax savings.

Apply

7. Using $r = 0.06/12 = 0.005$ and $n = 12(5) = 60$,

$$L = PMT\left[\frac{1 - (1+r)^{-n}}{r}\right]$$

$$L = 200\left[\frac{1 - (1+0.005)^{-60}}{0.005}\right] = \$10,345.11.$$

9. Using $r = 0.08/4 = 0.02$ and $n = 4(5) = 20$,

$$L = 250\left[\frac{1-(1+0.02)^{-20}}{0.02}\right] = \$4087.86$$

11. Using $r = 0.04/4 = 0.01$ and $n = 4(10) = 40$,

$$20{,}000 = PMT\left[\frac{1-(1+0.01)^{-40}}{0.01}\right]$$

$$20{,}000 = PMT(32.834686)$$

$$PMT = \$609.11 \text{ per quarter}.$$

13. Using $r = 0.08/4 = 0.02$ and $n = 4(20) = 80$,

$$120{,}000 = PMT\left[\frac{1-(1+0.02)^{-80}}{0.02}\right]$$

$$120{,}000 = PMT(39.744514)$$

$$PMT = \$3019.28 \text{ per quarter}.$$

15. Using $r = 0.09/12 = 0.0075$,

$$20{,}000 = 200\left[\frac{1-(1+0.0075)^{-n}}{0.0075}\right]$$

$$100 = \frac{1-(1+0.0075)^{-n}}{0.0075}$$

$$0.75 = 1 - 1.0075^{-n}$$

$$1.0075^{-n} = 0.25$$

$$-n\ln 1.0075 = \ln 0.25$$

$$n = \frac{\ln 0.25}{-\ln 1.0075} \approx 185.53.$$

Therefore, it takes 186 monthly payments or 15.5 years.

17. Using $r = 0.084/4 = 0.021$,

$$180{,}000 = 5000\left[\frac{1-(1+0.021)^{-n}}{0.021}\right]$$

$$36 = \frac{1-(1+0.021)^{-n}}{0.021}$$

$$0.756 = 1 - 1.021^{-n}$$

$$1.021^{-n} = 0.244$$

$$-n\ln(1.021) = \ln(0.244)$$

$$n = \frac{\ln(0.244)}{-\ln(1.021)} \approx 67.87.$$

Therefore, it takes 68 quarterly payments or 17 years.

Explore

19. Using $r = 0.105/12 = 0.00875$ and $n = 12(15) = 180$,

$$109{,}000 = PMT\left[\frac{1-(1+0.00875)^{-180}}{0.00875}\right]$$

$$109{,}000 = PMT(90.465078)$$

$$PMT = \$1204.88 \text{ per month}.$$

21. (a) Using $r = 0.132/12 = 0.011$ and $n = 12(4) = 48$,

$$8000 = PMT\left[\frac{1-(1+0.011)^{-48}}{0.011}\right]$$

$$8000 = PMT(37.137629)$$

$$PMT = \$215.41 \text{ per month}.$$

(b) $215.41 \times 48 = \$10{,}339.68$

(c) $10{,}339.68 - 8000 = \$2339.68$

23. (a) Using $r = 0.135/12 = 0.01125$ and $n = 12(4) = 48$,

$$5000 = PMT\left[\frac{1-(1+0.01125)^{-48}}{0.01125}\right]$$

$$5000 = PMT(36.932637)$$

$$PMT = \$135.38 \text{ per month}.$$

(b) Using $L = 5000$, $PMT = 250$, and $r = 0.01125$,

$$5000 = 250 \left[\frac{1 - (1 + 0.01125)^{-n}}{0.01125} \right]$$

$$20 = \frac{1 - (1 + 0.01125)^{-n}}{0.01125}$$

$$0.225 = 1 - 1.01125^{-n}$$

$$1.01125^{-n} = 0.775$$

$$-n \ln 1.01125 = \ln 0.775$$

$$n = \frac{\ln 0.775}{-\ln 1.01125} \approx 22.78.$$

Therefore, it takes 23 months or 1 year and 11 months.

(c) Cost of the original loan: $48 \times 135.38 = \$6498.24$
Cost with the new payments: $23 \times 250 = \$5750.00$
Savings $= 6498.24 - 5750.00 = \$748.24$
Note that the actual cost is slightly lower than this value since only 22.78 and not 23 payments are required.

25. Using $r = 0.096/12 = 0.008$ and $n = 12(30) = 360$,

$$175,000 = PMT \left[\frac{1 - (1 + 0.008)^{-360}}{0.008} \right]$$

$$175,000 = PMT(117.902287)$$

$$PMT = \$1484.28 \text{ per month.}$$

After 9 years, there are 21 years left so $n = 12(21) = 252$.

$$L = 1484.28 \left[\frac{1 - (1 + 0.008)^{-252}}{0.008} \right] = \$160,625.11$$

27. (a) For the first loan, using $r = 0.0723/12 = 0.006025$ and $n = 12(15) = 180$,

$$150,000 = PMT \left[\frac{1 - (1 + 0.006025)^{-180}}{0.006025} \right]$$

$$150,000 = PMT(109.680888)$$

$$PMT = \$1367.60 \text{ per month.}$$

For the second loan, using $r = 0.075/12 = 0.00625$ and $n = 12(30) = 360$,

$$150,000 = PMT \left[\frac{1 - (1 + 0.00625)^{-360}}{0.00625} \right]$$

$$150,000 = PMT(143.017627)$$

$$PMT = \$1048.82 \text{ per month.}$$

(b) For the first loan, the total payments are 1367.60 × 180 = $246,168.00.
For the second loan, the total payments are 1048.82 × 360 = $377,575.20.

(c) The first loan is better because it has lower total payments.

29. (a) Assuming that you are paying the points in this first year, we calculate the first year cost of ownership as follows:

The amount of the mortgage will be 180,000 – 15,000 = $165,000.
Using $n = 30 \times 12 = 360$ and $r = 0.081/12 = 0.00675$, the mortgage payments on the condo will be

$$165,000 = PMT \left[\frac{1 - (1 + 0.00675)^{-360}}{0.00675} \right]$$

$$165,000 = PMT(134.998733)$$

$$PMT = \$1222.23 \text{ per month.}$$

After one year, the balance of the mortgage has decreased.
Since there are 29 years left on the mortgage, use $n = 29 \times 12 = 348$,
$r = 0.00675$, and PMT = $1222.23. This gives

$$L = 1222.23 \left[\frac{1 - (1.00675)^{-348}}{0.00675} \right] = \$163,648.27.$$

Therefore, the mortgage decreased by $165,000 – $163,648.27 = $1351.73.

The cost of points for the loan are $165,000 \times 0.02 = \$3300$.

Using a 20% tax rate, the monthly tax saving is estimated at

$$0.20 \times 1222.23 = \$244.45.$$

Therefore, the first year costs of the loan are estimated at

$$\text{Cost} = \text{Monthly Payments} + \text{Points} - \text{Tax Savings} - \text{Decrease in Mortgage}$$
$$= (12 \times 1222.23) + 3300 - (12 \times 244.45) - 1351.73$$
$$= \$13,681.63.$$

One year of renting costs $1650 \times 12 = \$19,800$.

Therefore, the first year cost of buying is less than another year of renting.

(b) Assuming that you are paying the points in this first year. we calculate the first year cost of ownership as follows:

The amount of the mortgage will be $180,000 - 36,000 = \$144,000$
Using $n = 30 \times 12 = 360$ and $r = 0.0795/12 = 0.006625$, the mortgage payments on the condo will be

$$144,000 = PMT\left[\frac{1-(1+0.006625)^{-360}}{0.006625}\right]$$

$$144,000 = PMT(136.933408)$$

$$PMT = \$1051.61 \text{ per month.}$$

After one year, the balance of the mortgage has decreased.

Since there are 29 years left on the mortgage, use $n = 29 \times 12 = 348$,

$r = 0.006625$, and PMT = $\$1051.61$. This gives

$$L = 1051.61\left[\frac{1-(1.006625)^{-348}}{0.006625}\right] = \$142,785.63.$$

Therefore, the mortgage decreased by $\$144,000 - \$142,785.63 = \$1214.37$.

The cost of points for the loan are $144,000 \times 0.015 = \$2160$.

Using a 20% tax rate, the monthly tax saving is estimated at
$$0.20 \times 1051.61 = \$210.32.$$

Therefore, the first year costs of the loan are estimated at

Cost = Monthly Payments + Points – Tax Savings – Decrease in Mortgage
$$= (12 \times 1051.61) + 2160 - (12 \times 210.32) - 1214.37$$
$$= \$11,041.05.$$

Thus, the first year costs are lower but you have to make a much larger downpayment.

An additional consideration is that over 30 years, the loans will cost differing amounts.

Using the original loan, the total payments plus the downpayment will be
$$1222.23 \times 360 + 15,000 = \$455,002.80.$$
Using the second loan, the total payments plus the downpayment will be
$$1051.61 \times 360 + 36,000 = \$414,579.60.$$

Therefore, if you use the larger downpayment, you could save a substantial amount of money over the life of the loan.

Review Section 9.1

1. (a) $\dfrac{31}{32} = 0.96875 = 96.875\%$
    (b) $31.25\% = 31.25 \div 100 = 0.3125$

2. $279.99 = C + 0.20 \times C$, therefore $C = 279.99 \div 1.2 = \$233.33$

3. $14,645 + 0.305(120,000 - 65,550) = \$31,252.25$

4. Bill's percentage decrease was $\dfrac{93 - 87}{93} \times 100 = 6.45\%$.

    Jill's percentage decrease was $\dfrac{88 - 83}{88} \times 100 = 5.68\%$.

    Bill had the greater percentage decrease.

Review Section 9.2

5. (a) $A = P(1 + rt) = 2000(1 + 0.084 \times 3) = \$2504$
    (b) $2000 = P(1 + 0.084 \times 3)$, $P = \$1597.44$

6. $A = 10,000(1 + 0.08 \times 3) = \$12,400$

7. Using $A = P(1 + rt)$

    $10,000 = 9300(1 + 2r)$

    $\dfrac{10,000}{9300} = 1 + 2r$

    $1.07527 = 1 + 2r$

    $2r = 0.07527$

    $r \approx 0.0376 = 3.76\%$.

8. Using $t = 3/12 = 0.25$, $A = 500(1 + 0.09 \times 0.25) = \$511.25$.

Review Section 9.3

9. (a) Using $r = 0.084 \div 12 = 0.007$ and $n = 12 \times 3 = 36$,
        $A = 2000(1 + 0.007)^{36} = \$2570.93$.
    (b) Using $r = 0.084 \div 12 = 0.007$ and $n = 12 \times 3 = 36$,

        $A = P(1 + r)^n$

        $2000 = P(1 + 0.007)^{36}$

        $2000 = 1.285467P$

        $P = \$1555.85$.

10. Using $r = 0.073 \div 365 = 0.0002$
$$12{,}000 = 5000(1 + 0.0002)^n$$
$$1.0002^n = \frac{12{,}000}{5000} = 2.4$$
$$n\ln(1.0002) = \ln(2.4)$$
$$n = \frac{\ln(2.4)}{\ln(1.0002)} = 4378 \text{ days} \approx 12 \text{ years}.$$

11. Using $n = 20 \times 12 = 240$,
$$A = P(1 + r)^n$$
$$24{,}000 = 3000(1 + r)^{240}$$
$$(1 + r)^{240} = 8$$
$$1 + r = 8^{1/240} \approx 1.0087$$
$$r \approx 0.0087 \text{ per month} \times 12 \times 100 \approx 10.4\%.$$

12. Using $r = 0.06 \div 12 = 0.005$, $r_{eff} = (1 + 0.005)^{12} - 1 \approx 0.0617 \approx 6.17\%$.

13. Using $r = 0.06 \div 4 = 0.015$ and $n = 3 \times 4 = 12$,
$$1500 = P(1 + 0.015)^{12}$$
$$1500 = 1.195618P$$
$$P = \frac{1500}{1.195618} \approx \$1254.58.$$

14. Using $r = 0.072 \div 12 = 0.006$ and $n = 12 \times 12 = 144$,
$$A = 1000(1 + 0.006)^{144} = \$2366.51.$$

15. In the first eleven day period $P = 2500.64$, $r = 0.0002164$, and $n = 11$. The balance is
$A = 2500.64(1 + 0.0002164)^{11} = \$2506.60$ and the interest is $I = 2506.60 - 2500.64 = \$5.96$.

In the following sixteen day period, $P = 2506.60 - 2200 = \$306.60$ and $n = 16$. The balance is $A = 306.60(1 + 0.0002164)^{16} = \$307.66$ and the interest is $I = 307.66 - 306.60 = \$1.06$.

In the final three day period, $P = 307.66 + 621.67 = \$929.33$ and $n = 3$. The balance is $A = 929.33(1 + 0.0002164)^3 = \$929.94$ and the interest is $I = 929.94 - 929.33 = \$0.60$.

Thus, the total finance charge for the period is $\$5.96 + \$1.06 + \$0.60 = \$7.63$.

Review Section 9.4

16. Using $n = 12 \times 3 = 36$ and $r = 0.084 \div 12 = 0.007$,

$$S = 200\left[\frac{(1+0.007)^{36} - 1}{0.007}\right] = \$8156.20.$$

17. Using $n = 12 \times 3 = 36$ and $r = 0.084 \div 12 = 0.007$,

$$9000 = PMT\left[\frac{1.007^{36} - 1}{0.007}\right]$$

$$9000 = 40.781003(PMT)$$

$$PMT = \frac{9000}{40.781003} = \$220.69.$$

18. Using $r = 0.07 \div 12 \approx 0.00583333$,

$$200{,}000 = 500\left[\frac{1.00583333^{n} - 1}{0.00583333}\right]$$

$$1.00583333^{n} - 1 = \frac{200{,}000}{500} \times 0.00583333 = 2.333333$$

$$1.00583333^{n} = 3.333333$$

$$n\ln(1.00583333) = \ln(3.333333)$$

$$n = \frac{\ln(3.333333)}{\ln(1.00583333)} \approx 207 \text{ months} = 17\,\text{years, } 3 \text{ months.}$$

19. Using $n = 32 \times 12 = 384$ and $r = 0.0765 \div 12 = 0.006375$,

$$500{,}000 = PMT\left[\frac{1.006375^{384} - 1}{0.006375}\right]$$

$$500{,}000 = 1643.2439(PMT)$$

$$PMT = \frac{500{,}000}{1643.2439} = \$304.28.$$

20. Using $r = 0.093 \div 12 = 0.00775$,

$$1,000,000 = 400\left[\frac{1.00775^n - 1}{0.00775}\right]$$

$$1.00775^n - 1 = \frac{1,000,000}{400} \times 0.00775 = 19.375$$

$$1.00775^n = 20.375$$

$$n\ln(1.00775) = \ln(20.375)$$

$$n = \frac{\ln(20.375)}{\ln(1.00775)} \approx 391 \text{ months} = 32 \text{ years, } 7 \text{ months.}$$

Total deposits $= 391 \times 400 = \$156,400$

Review Section 9.5

21. Using $r = 0.084 \div 12 = 0.007$ and $n = 12 \times 3 = 36$,

$$L = 200\left[\frac{1 - 1.007^{-36}}{0.007}\right] = \$6344.93.$$

22. Using $r = 0.084 \div 12 = 0.007$ and $n = 12 \times 3 = 36$,

$$9000 = PMT\left[\frac{1 - 1.007^{-36}}{0.007}\right]$$

$$9000 = 31.724659(PMT)$$

$$PMT = \frac{9000}{31.724659} \approx \$283.69.$$

23. Using $r = 0.07 \div 12 = 0.00583333$,

$$200,000 = 2000\left[\frac{1 - 1.00583333^{-n}}{0.00583333}\right]$$

$$1 - 1.00583333^{-n} = \frac{200,000}{2000} \times 0.00583333 = 0.583333$$

$$1.00583333^{-n} = 1 - 0.583333 = 0.416667$$

$$-n\ln(1.00583333) = \ln(0.416667)$$

$$n = \frac{\ln(0.416667)}{-\ln(1.00583333)} \approx 151 \text{ months} = 12 \text{ years, } 7 \text{ months.}$$

24. Using $r = 0.072 \div 12 = 0.006$ and $n = 12 \times 30 = 360$, we find the original loan payments are

$$150{,}000 = PMT\left[\frac{1-1.006^{-360}}{0.006}\right]$$

$$150{,}000 = 147.321357(PMT)$$

$$PMT = \frac{150{,}000}{147.321357} \approx \$1018.18.$$

If we increase the payments by $150, the new payments are $1168.18. Using this value in the loan formula, we have

$$150{,}000 = 1168.18\left[\frac{1-1.006^{-n}}{0.006}\right]$$

$$1-1.006^{-n} = \frac{150{,}000}{1168.18} \times 0.006 = 0.770429$$

$$1.006^{-n} = 1-0.770429 = 0.229571$$

$$-n\ln(1.006) = \ln(0.229571)$$

$$n = \frac{\ln(0.229571)}{-\ln(1.006)} \approx 246 \text{ months} = 20 \text{ years, } 6 \text{ months.}$$

25.   (a)   The loan is $15{,}000 - 0.20(15{,}000) = \$12{,}000$.

Using $r = 0.114/12 = 0.0095$ and $n = 12 \times 3 = 36$,

$$12{,}000 = PMT\left[\frac{1-1.0095^{-36}}{0.0095}\right]$$

$$12{,}000 = 30.368860(PMT)$$

$$PMT = \frac{12{,}000}{30.368860} \approx \$395.14.$$

   (b)   The total amount paid for the car was $3000 + 36 \times 395.14 = \$17{,}225.04$.

   (c)   Interest $= 17{,}225.04 - 15{,}000 = \$2225.04$.

26.   There are two years left on the loan so $n = 24$. Using $r = 0.09 \div 12 = 0.0075$,

$$L = 207.58\left[\frac{1-1.0075^{-24}}{0.0075}\right] = \$4543.75.$$

27. Assuming that you are paying the points in this first year. we calculate the first year cost of ownership as follows:

The amount of the mortgage will be $150,000 - 10,000 = \$140,000$.
Using $n = 30 \times 12 = 360$ and $r = 0.084/12 = 0.007$, the mortgage payments on the condo will be

$$140,000 = PMT\left[\frac{1-(1+0.007)^{-360}}{0.007}\right]$$

$$140,000 = PMT(131.261561)$$

$$PMT = \$1066.57 \text{ per month.}$$

After one year, the balance of the mortgage has decreased.

Since there are 29 years left on the mortgage, use $n = 29 \times 12 = 348$,

$r = 0.007$, and PMT = \$1066.57, giving $L = 1066.57\left[\frac{1-(1.007)^{-348}}{0.007}\right] = \$138,919.83$.

Therefore, the mortgage decreased by $\$140,000 - \$138,919.83 = \$1080.17$.

The cost of points for the loan are $140,000 \times 0.02 = \$2800$.

Using a 20% tax rate, the monthly tax saving is estimated at $0.20 \times 1066.57 = \$213.31$.

Therefore, the first year costs of the loan are estimated at

Cost = Monthly Payments + Points - Tax Savings - Decrease in Mortgage
$$= (12 \times 1066.57) + 2800 - (12 \times 213.31) - 1080.17$$
$$= \$11,958.95.$$

One year of renting costs $1500 \times 12 = \$18,000$.

Therefore, the first year cost of buying is less than another year of renting.

CHAPTER 9   TEST

1. The price after the discount is $P = 24.95 - 0.25 \times 24.95 = \$18.71$. Adding the sales tax, the actual price is $A = 18.71 + 0.065 \times 18.71 = \$19.93$.

2. (a) \$1806.25   (b) \$1750

3. The percent increase is given by $\dfrac{3000-2250}{2250} \times 100 = \dfrac{750}{2250} \times 100 = 33.33\%$.

4. $1,669,090.91

5. Using the formula $A = P(1 + rt)$ with $P = 1\text{-}804.85$, $A = 14,700$ and $t = 5$, we have

$$14,700 = 10,804.85(1 + 5r)$$

$$1 + 5r = \frac{14,700}{10804.85} \approx 1.36$$

$$5r = 0.36$$

$$r = \frac{0.36}{5} = 0.072 \text{ or } 7.2\%.$$

6. Using $n = 12$ and $r = 0.10/12 = 0.0083333$, $A = 1000(1 + 0.0083333)^{12} = \$1104.71$.

7. Using the compound interest formula, $A = P(1 + r)^n$, with $A = 25,000$,
$r = \dfrac{0.081}{12} = 0.00675$ and $n = 18 \times 12 = 216$, we have

$$25,000 = P(1.00675)^{216}$$

$$P = \frac{25,000}{1.00675^{216}} = \$5846.10.$$

8. (a) Using $r = 0.08/4 = 0.02$ and $n = 10 \times 4 = 40$,

$$S = 400\left[\frac{(1 + 0.02)^{40} - 1}{0.02}\right] = \$24,160.79.$$

   (b) Amount deposited $= 400 \times 40 = \$16,000$.
        Interest $= 24,160.79 - 16,000 = \$8,160.79$.

   (c) Using $r = 0.08/4 = 0.02$ and $n = 4(25) = 100$,
        $A = 24,160.79(1 + 0.02)^{100} = \$175,036.37$.

9. Using $A = P(1 + r)^n$ with $P = 1,000,000$, $A = 5,000,000$ and $r = \dfrac{0.06}{4} = 0.015$, we have

$$5,000,000 = 1,000,000(1.015)^n$$

$$1.015^n = 5$$

$$n\ln(1.015) = \ln(5)$$

$$n = \frac{\ln 5}{\ln 1.015} \approx 109 \text{ quarters } = 27.25 \text{ years}.$$

10. About 18.85 years

11. Using the formula $L = PMT\left[\dfrac{1-(1+r)^{-n}}{r}\right]$ with $PMT = \$926.21$, $r = \dfrac{0.097}{12} \approx 0.0080833$

and $n = 12 \times (30 - 9) = 252$, we have

$$L = 926.21\left[\dfrac{1-(1+0.0080833)^{-252}}{0.0080833}\right] = \$99,516.29.$$

12. $r_{eff} = \left(1+\dfrac{0.059}{4}\right)^4 - 1 \approx 0.0603$ or 6.03%

13. Using the formula $L = PMT\left[\dfrac{1-(1+r)^{-n}}{r}\right]$ with $L = 150,000$, $PMT = 2500$ and

$r = \dfrac{0.072}{12} = 0.006$, we have

$$150,000 = 2500\left[\dfrac{1-(1.006)^{-n}}{0.006}\right]$$

$$\dfrac{1-(1.006)^{-n}}{0.006} = \dfrac{150,000}{2500} = 60$$

$$1-1.006^{-n} = 60 \times 0.006 = 0.0 = 0.36$$

$$1.006^{-n} = 1-0.36 = 0.64$$

$$-n\ln(1.006) = \ln(0.64)$$

$$n = \dfrac{\ln(0.64)}{-\ln(1.006)} \approx 75 \text{ months} = 6 \text{ years, } 3 \text{ months.}$$

14. Assuming that you are paying the points in this first year. we calculate the first year cost of ownership as follows:

The amount of the mortgage will be $210,000 - 30,000 = \$180,000$.
Using $n = 30 \times 12 = 360$ and $r = 0.081/12 = 0.00675$, the mortgage payments on the condo will be

$$180,000 = PMT\left[\dfrac{1-(1+0.00675)^{-360}}{0.00675}\right]$$

$$180,000 = PMT(134.998733)$$

$$PMT = \$1333.35 \text{ per month.}$$

After one year, the balance of the mortgage has decreased.

Since there are 29 years left on the mortgage, use $n = 29 \times 12 = 348$,

$r = 0.000675$, and PMT = $1333.35, giving

$$L = 1333.35\left[\frac{1-(1.00675)^{-348}}{0.00675}\right] = \$178{,}525.92.$$

Therefore, the mortgage decreased by $180,000 – $178,525.92 = $1474.08

The cost of points for the loan are $180{,}000 \times 0.02 = \$3600$.

Using a 20% tax rate, the monthly tax saving is estimated at
$0.20 \times 1333.35 = \$266.67$.

Therefore, the first year costs of the loan are estimated at

Cost = Monthly Payments + Points – Tax Savings – Decrease in Mortgage
$$= (12 \times 1333.35) + 3600 - (12 \times 266.67) - 1474.08$$
$$= \$14{,}926.04.$$

One year of renting costs $1800 \times 12 = \$21{,}600$.

Therefore, the first year cost of buying is less than another year of renting.

15. In the first eleven day period $P = 7159.68$, $r = 0.00030822$, and $n = 10$. The balance is

$$A = 7159.68(1 + 0.00030822)^{10} = \$7181.78$$

and the interest is $I = 7181.78 - 7159.68 = \$22.10$.

In the following seventeen day period, $P = 7181.78 - 1850 = \$5331.78$ and $n = 17$. The balance is

$$A = 5331.78(1 + 0.00030822)^{17} = \$5359.78$$

and the interest is $I = 5359.78 - 5331.78 = \$28.01$.

In the final three day period, $P = 5359.78 + 320.67 = 5680.45$ and $n = 3$. The balance is

$$A = 5680.45(1 + 0.00030822)^{3} = \$5685.71$$

and the interest is $I = 5685.71 - 5680.45 = \$5.25$.

Thus, the total finance charge for the period is $22.10 + $28.01 + $5.25 = $55.36.

CHAPTER 10   MATH FROM OTHER VISTAS

In the nine previous chapters, we investigated the math that could follow the arithmetic, algebra, and geometry studied by the liberal arts student. That does not imply that there is no more mathematics to study. The study of math is limitless. In this chapter, we will introduce other areas of mathematics that would complement the previous studies of the liberal arts student.

In Sections 10.1 and 10.2, we will examine what is considered the culmination of arithmetic, algebra, geometry, and trigonometry – Calculus. In Section 10.3, we will look at an ancient triangular array of numbers that has some interesting applications. In Section 10.4, we will look at some of the methods used in determining the winner of an election. In Section 10.5, we will investigate the way a representative body of leaders can be selected in a democratic system. Finally, in Section 10.6, we will examine a technique to maximize profits for a business.

SECTION 10.1 DIFFERENTIAL CALCULUS

Section 10.1 introduces the concept of differential calculus from the perspective of the slope of the tangent line to a curve. You are also presented with the standard formulas for taking derivatives of polynomials.

Explain

1.   A tangent line to a curve is a line that, in a region very near to a point, touches the curve at only that point.

3.

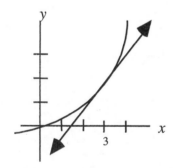

5.   If $f(x)$ is a constant function, its derivative is equal to zero and the graph of $f(x)$ is a horizontal line.

Apply

7.   (a) B, D, F (b) A, E (c) C, G

9.   At $x = 2$, the derivative has a value between 0.5 and 2.

11. 2

13. $-2x$

15. $9x^2 - 5$

$$\frac{dy}{dx} = 3 \cdot 3x^{3-1} - 5 - 0 = 9x^2 - 5$$

17. $4x^3 + 3x^2 + 2x + 1$

$$\frac{dy}{dx} = 4x^{4-1} + 3x^{3-1} + 2x^{2-1} + 1 + 0 = 4x^3 + 3x^2 + 2x + 1$$

19. $20x^4$

$$\frac{dy}{dx} = 4 \cdot 5x^{5-1} = 20x^4$$

21. $20x^4 - 10x$

$$\frac{dy}{dx} = 4 \cdot 5x^{5-1} - 5 \cdot 2x^{2-1} = 20x^4 - 10x$$

23. $5x^4 - 6x + 7$

$$\frac{dy}{dx} = 5x^{5-1} - 3 \cdot 2x^{2-1} + 7 + 0 = 5x^4 - 6x + 7$$

25. (a)   The graph does not have any relative maximum or minimum points.

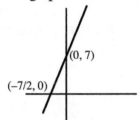

(0, 7)

(−7/2, 0)

(b) $\dfrac{dy}{dx} = 3$. The derivative is never equal to zero. It is always 3.

27. (a) The graph has a relative maximum.

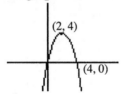

(2, 4)

(4, 0)

(b) $\dfrac{dy}{dx} = -2x + 4$

If $-2x + 4 = 0$, then $x = 2$.

(c)   maximum

29. (a) By plotting points, we get the following graph.

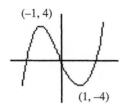

(b) $\dfrac{dy}{dx} = 6x^2 - 6$

(c)
$$6x^2 - 6 = 0$$
$$6(x^2 - 1) = 0$$
$$6(x - 1)(x + 1) = 0$$
$$x = 1 \text{ or } -1$$

(d) $x = -1$, max; $x = 1$, min

Explore

31. (a) 42 in. (b) 18 in.
$$\frac{dh}{dx} = \frac{-1}{98}(2x) + \frac{6}{7} = \frac{-x}{49} + \frac{6}{7}$$

Setting the derivative equal to zero, we get the following.

$$\frac{-x}{49} + \frac{6}{7} = 0$$
$$\frac{-x}{49} = -\frac{6}{7}$$
$$-x = -42$$
$$x = 42$$

If $x = 42$, then $h = \dfrac{-1}{98}(42)^2 + \dfrac{6}{7}(42) = 18$.

33. 150 ft by 300 ft; area = 45,000 sq ft

Since the amount of fencing is 1200 ft,
$$4x + 2y = 1200$$
$$2y = 1200 - 4x$$
$$y = 600 - 2x.$$

By substituting this expression for y into the formula for the area of the region, we can then use the derivative to find the maximum area.

$$A = xy$$

$$= x(600 - 2x)$$

$$= 600x - 2x^2$$

$$\frac{dA}{dx} = 600 - 4x$$

$$600 - 4x = 0$$

$$600 = 4x$$

$$150 = x$$

Thus, $y = 600 - 2x = 600 - 2(150) = 300$.

$$A = xy = 150(300) = 45,000.$$

35. 2 ft; 4 ft; $2\frac{2}{3}$ ft, $V = \frac{64}{3}$ cu ft

If x is the width of the box, the length is $2x$. The relationship between x and the height of the box (h) can be found from the surface area equation for the total plywood.

$$48 = 4x^2 + 4xh + 2xh$$

$$48 = 4x^2 + 6xh$$

$$48 - 4x^2 = 6xh$$

$$\frac{48 - 4x^2}{6x} = h$$

$$\frac{24 - 2x^2}{3x} = h$$

By substituting this expression for h into the formula for the volume of the box, we can use the derivative to find the maximum volume.

$$V = 2x^2 h$$

$$= 2x^2 \left(\frac{24 - 2x^2}{3x} \right)$$

$$= 16x - \frac{4}{3}x^3$$

$$\frac{dV}{dx} = 16 - 4x^2$$

If $16 - 4x^2 = 0$, $x = 2$.

Thus, the length is 4, $h = \dfrac{24 - 2x^2}{3x} = \dfrac{24 - 2(2)^2}{3(2)} = \dfrac{16}{6} = 2\frac{2}{3}$,

and $V = 2x^2 h = 2(4)\left(2\frac{2}{3}\right) = \frac{64}{3}$.

37. (a) 20,000   (b) $16.20

x = the number of shoes
C = the average cost per shoe

$$C = 80,000x^{-1} + 0.0002x + 8.2$$

$$\frac{dC}{dx} = -80,000x^{-2} + 0.0002$$

Setting the derivative equal to zero and solving for x, we get

$$-80,000x^{-2} + 0.0002 = 0$$

$$\frac{-80,000}{x^2} + 0.0002 = 0$$

$$0.0002 = \frac{80,000}{x^2}$$

$$0.0002x^2 = 80,000$$

$$x^2 = 400,000,000$$

$$x = 20,000$$

and $C = \dfrac{80,000}{20,000} + 0.0002(20,000) + 8.2 = \$16.20.$

SECTION 10.2 INTEGRAL CALCULUS

Integration is one of the two main branches of calculus. One of the applications of integration is to determine the area bounded by a curve. In Section 10.2, we use rectangles to approximating such areas. We also use the definite integral to find the exact area under a curve.

Explain

1. Integration is the process used to find the area of a region in a plane.

3. Lower rectangles are rectangles which have their lengths determined by the minimum value of the function within a given interval. The area approximated by the lower rectangles will be less than the actual area of a region. However, as the number of rectangles increases, lower rectangles give an area that is close to the actual area under the curve.

Apply

5. Using three rectangles, the width of each rectangle is 4. The length of each mid-point rectangle is the y-value for each x-value.

x	y
2	20
6	12
10	4

$A = 4(20) + 4(12) + 4(4) = 144$

7. Using eight rectangles, the width of each rectangle is 2. The length of each mid-point rectangle is the y-value for each x-value.

x	y
1	6
3	10
5	14
7	18

$A = 2(6) + 2(10) + 2(14) + 2(18) = 96$

9. Using the Fundamental Theorem of Calculus, we have

$$\int_0^5 \left(x^2\right)dx = \left(\frac{x^3}{3}\right)\Bigg|_0^5$$
$$= \frac{5^3}{3} - 0$$
$$= \frac{125}{3}.$$

11. The function intersects the y axis at -6 and 6. Using these as the limits in the Fundamental Theorem of Calculus, we have

$$\int_{-6}^6 \left(36 - x^2\right)dx = \left(36x - \frac{x^3}{3}\right)\Bigg|_{-6}^6$$
$$= \left(36 \times 6 - \frac{6^3}{3}\right) - \left(36 \times (-6) - \frac{(-6)^3}{3}\right)$$
$$= (216 - 72) - (-216 + 72)$$
$$= 288.$$

Explore

13. $A = 25(90) + 25(120) + 25(90) + 25(55) + 25(35) = 9750$ sq ft.

15. (a) Graph of region bounded by $y = \frac{1}{x}$, $x = 1$, $x = 4$, and the x axis.

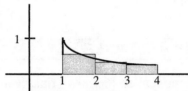

(b) Using three rectangles, the width of each rectangle is 1. The length of each mid-point rectangle is the y-value for each x-value.

x	y
1.5	0.667
2.5	0.4
3.5	0.286

$A = 1(0.667) + 1(0.4) + 1(0.286) \approx 1.35$

17. (a) 125.66 sq cm

$A = \pi ab$

$\quad = \pi(8)(5) = 40\pi \approx 125.66$

(b) 124.56 sq cm

The widths of the midpoint rectangles are 2 cm.

Solving for y in the equation of the ellipse, we get, $y = \frac{5}{8}\sqrt{64 - x^2}$.

Thus, the lengths of the rectangles are as follows:

At $x = -7$, $l_1 = 2\left(\frac{5}{8}\sqrt{64 - (-7)^2}\right) = 4.841$

At $x = -5$, $l_2 = 2\left(\frac{5}{8}\sqrt{64 - (-5)^2}\right) = 7.806$

At $x = -3$, $l_3 = 2\left(\frac{5}{8}\sqrt{64 - (-3)^2}\right) = 9.270$

At $x = -1$, $l_4 = 2\left(\frac{5}{8}\sqrt{64 - (-1)^2}\right) = 9.922$

At $x = 1$, $l_5 = 2\left(\frac{5}{8}\sqrt{64 - (1)^2}\right) = 9.922$

At $x = 3$, $l_6 = 2\left(\frac{5}{8}\sqrt{64 - (3)^2}\right) = 9.270$

At $x = 5$, $l_7 = 2\left(\frac{5}{8}\sqrt{64 - (5)^2}\right) = 7.806$

At $x = 7$, $l_8 = 2\left(\frac{5}{8}\sqrt{64 - (7)^2}\right) = 4.841$.

The total area of the rectangles is:

$$A = 2(4.841) + 2(7.806) + 2(9.270) + 2(9.222) + 2(9.222) +$$
$$2(9.270) + 2(7.806) + 2(4.841) = 124.56.$$

(c) The difference in the results is about 1.1 sq cm.

SECTION 10.3 THE PASCAL–YANGHUI TRIANGLE

In this section, you are introduced to what is commonly called Pascal's Triangle after the French mathematician, Blaise Pascal, 1665. However, there is evidence that Chinese mathematicians described this triangular pattern some 500 years before Pascal. Since in China it is called the Yanghui Triangle, the triangle will be referred to as the Pascal–Yanghui Triangle in the text.

This section will show you how to use the triangle to expand binomials and determine combinations and to use modified Pascal–Yanghui triangles to expand polynomials and solve "dice" problems. By using such a triangle, you will be able to solve a problem such as, "*If five standard dice are tossed, what is the number of ways that the sum of the dice is 16?*"

Explain

1. There is evidence that Chinese mathematicians described the triangular pattern some 500 years before Pascal. Since in China it is called the Yanghui Triangle, the triangle is referred to as the Pascal–Yanghui Triangle in the text.

3. Each row starts and ends with the number one and the other entries in each successive row are obtained by adding the two numbers above it.

5. It gives you the coefficients of each term in the expansion of $(1 + x)^6$.

7. It gives the coefficients of each term in the expansion of $\left(1 + x + x^2 + x^3\right)^3$.

9. The entries of the sixth row gives:
    $C_{6,0} = 1,\ C_{6,1} = 6,\ C_{6,2} = 15,\ C_{6,3} = 20,\ C_{6,4} = 15,\ C_{6,5} = 6,\ C_{6,6} = 1$

11. The fourth row of Triangle–3 gives the number of ways to get a sum of 4, 5, 6, 7, 8, 9, 10, 11, and 12 with four "three-sided dice."

Apply

13. $1 + 7x + 21x^2 + 35x^3 + 35x^4 + 21x^5 + 7x^6 + x^7$
    Use the seventh row of the Pascal–Yanghui Triangle for the coefficients. The exponents of x start at 0 for the first term and increase to 7 at the last term.

15. $1 + 6x + 21x^2 + 50x^3 + 90x^4 + 126x^5 + 141x^6 + 126x^7 + 90x^8 + 50x^9$
$+21x^{10} + 6x^{11} + x^{12}$

Use the sixth row of Triangle–3 for the coefficients. The exponents of x start at 0 for the first term and increase to 12 at the last term.

17. $1 + 4x + 10x^2 + 20x^3 + 31x^4 + 40x^5 + 44x^6 + 40x^7 + 31x^8 + 20x^9 + 10x^{10} + 4x^{11} + x^{12}$

Use the fourth row of Triangle–4 for the coefficients. The exponents of x start at 0 for the first term and increase to 12 at the last term.

19. 9
In the ninth row of the Pascal–Yanghui Triangle, the first entry gives the number of ways to get a sum of 9. So, the ninth entry (9) gives the number of ways to get a sum of 17.

21. 15
In the fifth row of Triangle–3, the first entry gives the number of ways to get a sum of 5. So, the ninth entry (15) gives the number of ways to get a sum of 13.

23. 31
In the fourth row of Triangle–4, the first entry gives the number of ways to get a sum of 4. So, the tenth entry (31) gives the number of ways to get a sum of 12.

Explore

25. (a) 1, 3, 6, 10, 15, 18, 19, 18, 15, 10, 6, 3, 1
1, 4, 10, 20, 35, 52, 68, 80, 85, 80, 68, 52, 35, 20, 10, 4, 1
1, 5, 15, 35, 70, 121, 185, 255, 320, 365, 381, 365, 320, 255, 185, 121, 70, 35, 15, 5, 1
Each entry in successive rows are obtained by adding five items from the previous row.

(b) $1 + 3x + 6x^2 + 10x^3 + 15x^4 + 18x^5 + 19x^6 + 18x^7 + 15x^8 + 10x^9$
$+6x^{10} + 3x^{11} + x^{12}$

Use the third row of Triangle–5 for the coefficients. The exponents of x start at 0 for the first term and increase to 12 at the last term.

(c) $1 + 4x + 10x^2 + 20x^3 + 35x^4 + 52x^5 + 68x^6 + 80x^7 + 85x^8 + 80x^9 + 68x^{10}$
$+52x^{11} + 35x^{12} + 20x^{13} + 10x^{14} + 4x^{15} + x^{16}$

Use the fourth row of Triangle–5 for the coefficients. The exponents of x start at 0 for the first term and increase to 16 at the last term.

(d) 80 In the fourth row of Triangle–5, the first entry gives the number of ways to get a sum of 4. So, the tenth entry (80) gives the number of ways to get a sum of 13.

(e) 320 In the fifth row of Triangle–5, the first entry gives the number of ways to get a sum of 5. So, the ninth entry (320) gives the number of ways to get a sum of 13.

In this section, you examine different voting systems. The system that most people are familiar with is that the person with the most votes wins. However, if there are 5 people in an election and the winner receives 30% of the votes, it means that 70% of the voters voted against the winning candidate. This section examines several alternatives: Plurality, Plurality with Elimination, Borda Count, and Pairwise Comparison.

Explain

1. The main disadvantage of the plurality method is that it does not use anything other than first choices into consideration.

3. The main disadvantage of the Borda Count method is that a candidate who has no first place votes can be the winning candidate.

Apply

5.

	Ann	Brenda	Cathy	Dawn	Elle	Francesca	Gloria
The Time Machine	1	2	1	2	4	4	2
The Mysterious Island	4	4	3	1	1	3	3
The Secret Garden	2	3	4	3	3	2	1
Wind in the Willows	3	1	2	4	2	1	4

Plurality
The Time Machine has 2 first place votes
The Mysterious Island has 2 first place votes
The Secret Garden has 1 first place votes
The Wind in the Willows has 2 first place votes
Therefore, *The Time Machine*, *The Mysterious Island* and *The Wind in the Willows* are tied.

Plurality with Elimination
Removing *The Secret Garden* gives

	Ann	Brenda	Cathy	Dawn	Elle	Francesca	Gloria
The Time Machine	1	2	1	2	3	3	1
The Mysterious Island	3	3	3	1	1	2	2
Wind in the Willows	2	1	2	3	2	1	3

This results in *The Time Machine* winning with three first place votes.

Borda Count

	points
The Time Machine	$(2 \times 4) + (3 \times 3) + (0 \times 2) + (2 \times 1) = 19$
The Mysterious Island	$(2 \times 4) + (0 \times 3) + (3 \times 2) + (2 \times 1) = 16$
The Secret Garden	$(1 \times 4) + (2 \times 3) + (3 \times 2) + (1 \times 1) = 17$
Wind in the Willows	$(2 \times 4) + (2 \times 3) + (1 \times 2) + (2 \times 1) = 18$

Therefore, *The Time Machine* wins.

Pairwise Comparison

comparison	winner
Time v. *Mysterious*	*Time*
Time v. *Secret*	*Time*
Time v. *Wind*	*Time*
Mysterious v. *Secret*	*Secret*
Mysterious v. *Wind*	*Wind*
Secret v. *Wind*	*Wind*

Therefore, *The Time Machine* wins.

7.

	1200	200	1000	1500	100
1st place	A	A	B	C	B
2nd place	B	C	A	B	C
3rd place	C	B	C	A	A

Plurality: *A* wins with 1400.

Plurality with Elimination: *A* wins with 1400

Borda:

candidate	points
A	$(1400 \times 3) + (1000 \times 2) + (1600 \times 1) = 7800$
B	$(1100 \times 3) + (2700 \times 2) + (200 \times 1) = 8900$
C	$(1500 \times 3) + (300 \times 2) + (2200 \times 1) = 7300$

B wins

Pairwise

comparison	winner
A v. *B*	*B* wins 2600 to 1400
A v. *C*	*A* wins 2400 to 1600
B v. *C*	*B* wins 2300 to 1700

B wins.

Therefore, there is no clear winner.

Explore

9.

	13	10	8	7	2
Increase sales taxes	1	5	2	3	5
Increase income taxes	5	1	3	2	4
Decrease school funding	2	2	5	4	5
Decrease transportation funding	3	4	1	5	2
Decrease social service funding	4	3	2	1	1

The legislators agreed to use the Borda count method with the following weighting: first choice: 5 points, second choice 4 points, etc.

	points
sales tax	$(13 \times 5) + (8 \times 4) + (7 \times 3) + (0 \times 2) + (12 \times 1) = 130$
income tax	$(10 \times 5) + (7 \times 4) + (8 \times 3) + (2 \times 2) + (13 \times 1) = 119$
school	$(0 \times 5) + (23 \times 4) + (0 \times 3) + (7 \times 2) + (10 \times 1) = 116$
transportation	$(8 \times 5) + (2 \times 4) + (13 \times 3) + (10 \times 2) + (7 \times 1) = 114$
social services	$(9 \times 5) + (8 \times 4) + (10 \times 3) + (13 \times 2) + (0 \times 1) = 133$

First: Decrease social service funding
Second: Increase sales taxes
Third: Increase income taxes
Fourth: Decrease school funding
Fifth: Decrease transportation funding

11. First choice: 5 points, second choice: 5 points, third choice: 3 points, fourth choice: 2 points, and last choice: 1 point.

	points
sales tax	$(13 \times 6) + (8 \times 6) + (7 \times 3) + (0 \times 2) + (12 \times 1) = 159$
income tax	$(10 \times 6) + (7 \times 6) + (8 \times 3) + (2 \times 2) + (13 \times 1) = 143$
school	$(0 \times 6) + (23 \times 6) + (0 \times 3) + (7 \times 2) + (10 \times 1) = 162$
transportation	$(8 \times 6) + (2 \times 6) + (13 \times 3) + (10 \times 2) + (7 \times 1) = 126$
social services	$(9 \times 6) + (8 \times 6) + (10 \times 3) + (13 \times 2) + (0 \times 1) = 158$

First: Decrease school funding
Second: Increase sales taxes
Third: Decrease social service funding
Fourth: Increase income taxes
Fifth: Decrease transportation funding

SECTION 10.5 APPORTIONMENT

Apportionment is the process by which a given amount of a resource can be divided into parts. In particular, apportionment is used to assign congressional representation according to the population of each state. Preliminary to apportioning the representatives, for each county we find a quota. A **quota** is the number of seats that should be assigned based on proportional representation. For example, if California has 10% of the U. S. population, it should receive 10% of the votes in the House of Representatives. However, since there are 456 members of the House of Representative, California would be entitled to 45.6 representatives. Since the number of representatives assigned to a state must be a whole number, various methods have been devised in the attempt to reach an equitable apportionment while still assigning a whole number of representatives

In this section, you study three methods of apportionment — Hamilton's, Jefferson's and Webster's methods. As you learn these methods, you also learn that each of the methods has its own strengths and weaknesses.

Explain

1. Apportionment is the process by which a given amount of a resource can be divided into parts.

3. Jefferson's method apportions a resource by dividing the number in each group by a divisor and then rounding each quotient <u>down</u> to the nearest integer. If any of the resource remains unallocated, the divisor is decreased until all of the resource is allocated. In all cases, the largest possible divisor must be used.

5. The Alabama paradox occurs if the total amount of a resource is increased, the population remains the same, and the apportionment for a group decreases.

7. The quota rule states that the apportionment received by a group should be within one of its quota.

Apply

9.

Number of Seats Using Hamilton's Method

County	Population	Quota	Initial # of Seats	# Seats
Brooke	48,859	4.87	4	5
Hopkins	161,135	16.06	16	16
Isaac	87,194	8.69	8	9
Haley	596,270	59.42	59	59
Wallace	110,006	10.96	10	11
Totals	1,003,464	100.00	97	100

11.

Number of Seats Using Jefferson's Method

County	Population	Divisor = 10,035 Quota	# of Seats	Divisor = 9774 Pop/Divisor	# of Seats
Brooke	48,859	4.87	4	4.999	4
Hopkins	161,135	16.06	16	16.49	16
Isaac	87,194	8.69	8	8.92	8
Haley	596,270	59.42	59	61.01	61
Wallace	110,006	10.96	10	11.25	11
Totals	1,003,464	100.00	97		100

13.

Number of Seats Using Webster's Method

County	Population	Divisor = 16,248 Quota	# of Seats
Carlisle	110,993	6.83	7
Newport	441,946	27.20	27
Sonoma	113,229	6.97	7
Totals	666,168	41.00	41

Explore

15.

Number of Courses Using Hamilton's Method

Course	# Student	Quota	Initial # of Courses	# Courses
College Algebra	120	4.00	4	4
Calculus	80	2.67	2	3
Math for Business	235	7.83	7	8
Liberal Arts Math	165	5.50	5	5
Totals	600	20.00	18	20

17.

Number of Courses Using Jefferson's Method

Course	# Students	Divisor = 30		Divisor = 27	
		Quota	# of Courses	New quota	# of Seats
College Algebra	120	4.00	4	4.44	4
Calculus	80	2.67	2	2.96	2
Math for Business	235	7.83	7	8.70	8
Liberal Arts Math	165	5.50	5	6.11	6
Totals	600	20.00	18		20

19.

Number of Trucks Using Webster's Method

District	Population	Divisor = 544	
		Quota	# of Trucks
Bayview	25,456	46.80	47
Downtown	32,723	60.16	60
East End	27,568	50.69	51
Greenhaven	16,475	30.29	30
Ingleside	8,696	15.99	16
Riverview	11,458	21.07	21
Totals	122,376	225.00	225

21.

Number of Sites Using Hamilton's Method

District	Population	Quota	Initial # of Sites	# of Sites
1	90,000	90.91	90	91
2	166,000	167.68	167	168
3	102,000	103.03	103	103
4	131,000	132.32	132	132
5	147,000	148.48	148	148
6	156,000	157.58	157	158
Totals	792,000	800.00	797	800

23.

Number of Sites Using Jefferson's Method

District	Population	Pop/Divisor	# of Sites
1	90,000	91.19	91
2	166,000	168.19	168
3	102,000	103.34	103
4	131,000	132.73	132
5	147,000	148.94	148
6	156,000	158.05	158
Totals	792,000		800

25.

Number of Days Using Webster's Method

District	Voters	Voters/Divisor	# of Days
1	167,000	26.02	26
2	116,000	18.07	18
3	237,000	36.93	37
4	231,000	35.99	36
5	215,000	33.50	33
Totals	966,000		150

SECTION 10.6 LINEAR PROGRAMMING

Linear Programming is a mathematical technique for optimizing a linear function subject to one or more restrictions. The linear function to be optimized is called the **objective function**. The restrictions are called the **constraints** and are given as linear inequalities. The region described by the constraints is called the **feasible region**.

The method used to solve the linear programming problems in this section consists of three steps.

(1) Sketch the feasible region.
(2) Determine the vertices (corner points) of the feasible region.
(3) Optimize the objective function by substituting the coordinates of the vertices into the objective function.

Explain

1. The feasible region is the region that satisfies all the constraints.

3. To determine the maximum value in a linear programming problem,
   (i) Sketch the feasible region.
   (ii) Determine the vertices (corner points) of the feasible region.
   (iii) Maximize the objective function by substituting the coordinates of the vertices into the objective function. The maximum occurs at one or more of the vertices.

5. A constraint is a linear inequality that forms a restriction on the objective function.

Apply

7. Maximum $P = 75$ at $(10, 9)$.

x	y	$P = 3x + 5y$
0	0	0
12	0	36
10	9	75
0	14	70

9. Minimum $P = 27$ at $(0, 9)$.

x	y	$C = 12x + 3y$
9	0	108
3	2	42
0	9	27

11. Maximum $P = 288$ at $(14, 12)$.

x	y	$P = 12x + 10y$
2	11	134
5	9	150
11	9	222
14	12	288
11	14	272
5	14	200

13. Maximum $P = 116$ at $(9, 14)$. Minimum $P = 25$ at $(2, 3)$.

x	y	$P = 2x + 7y$
2	3	25
9	2	32
14	9	91
9	14	116
1	12	86

15. Maximum $P = 40$ at $(0, 8)$.

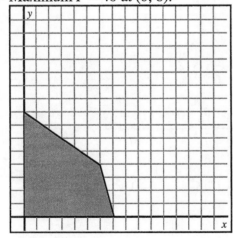

x	y	$P = 3x + 5y$
0	0	0
7	0	21
6	4	38
0	8	40

17. Maximum $P = 88$ at $(0, 8)$.

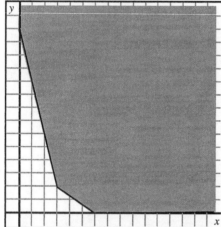

x	y	$P = 2x + 11y$
0	0	0
7	0	14
5	5	65
0	8	88

19. Minimum $C = 50$ at $(6, 4)$.

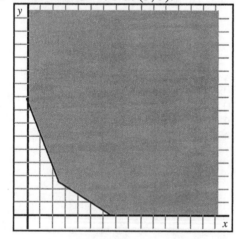

x	y	$C = 7x + 2y$
12	0	84
6	4	50
0	28	56

21. Minimum $C = 95$ at $(5, 5)$.

x	y	$P = 11x + 8y$
13.33	0	146.67
5	5	95
0	17.5	140

23. Minimum $P = 16$ at $(8, 0)$. Maximum $P = 56$ at $(4, 12)$.

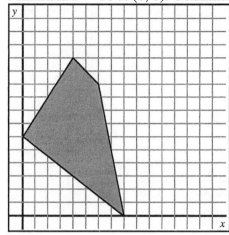

x	y	$P = 2x + 4y$
0	6	24
8	0	16
6	10	52
4	12	56

Explore

25. 60 grams of mix A and 100 grams of mix B gives 4.8 grams of fat.
    Let x = the number of grams of mix A and let y = the number of grams of B.
    Minimize $P = 0.03x + 0.03y$
    Subject to $0.25x + 0.10y \geq 25$
             $0.15x + 0.18y \geq 27$
             $x \geq 0, y \geq 0$

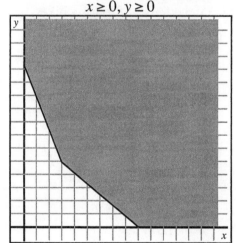

x	y	$P = 0.03x + 0.03y$
0	250	7.5
60	100	4.8
180	0	5.4

27. Use 400 pounds of each mix. Minimum cost is $6000.

Let x = the number of pounds of the first fertilizer and let y = the number of pounds of the second fertilizer.

Minimize $P = 9x + 6y$

Subject to
$$0.03x + 0.06y \geq 27$$
$$0.02x + 0.01y \geq 12$$
$$0.01x + 0.01y \geq 8$$
$$x \geq 0, y \geq 0$$

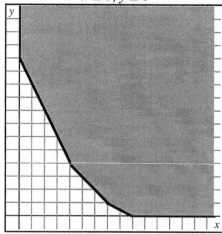

x	y	$P = 9x + 6y$
0	1200	7200
400	400	6000
700	100	6900
900	0	8100

29. 18 mountain bikes and 2 touring bikes. Maximum $P = \$1460$.

Let x = the number of mountain bikes and let y = the number of touring bikes.

Maximize $P = 70x + 100y$

Subject to
$$2x + 2y \leq 40$$
$$2x + 3y \leq 42$$
$$x \geq 0, y \geq 0$$

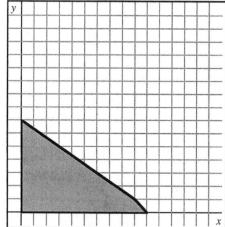

x	y	$P = 70x + 100y$
0	0	0
20	0	1400
18	2	1460
0	14	1400